MW01530652

PREVENTATIVE PROGRAMMING TECHNIQUES

AVOID AND CORRECT COMMON MISTAKES

THE CD-ROM WHICH ACCOMPANIES THE BOOK MAY BE USED ON A SINGLE PC ONLY. THE LICENSE DOES NOT PERMIT THE USE ON A NETWORK (OF ANY KIND). YOU FURTHER AGREE THAT THIS LICENSE GRANTS PERMISSION TO USE THE PRODUCTS CONTAINED HEREIN, BUT DOES NOT GIVE YOU RIGHT OF OWNERSHIP TO ANY OF THE CONTENT OR PRODUCT CONTAINED ON THIS CD-ROM. USE OF THIRD PARTY SOFTWARE CONTAINED ON THIS CD-ROM IS LIMITED TO AND SUBJECT TO LICENSING TERMS FOR THE RESPECTIVE PRODUCTS.

CHARLES RIVER MEDIA, INC. ("CRM") AND/OR ANYONE WHO HAS BEEN INVOLVED IN THE WRITING, CREATION, OR PRODUCTION OF THE ACCOMPANYING CODE ("THE SOFTWARE") OR THE THIRD PARTY PRODUCTS CONTAINED ON THE CD-ROM OR TEXTUAL MATERIAL IN THE BOOK, CANNOT AND DO NOT WARRANT THE PERFORMANCE OR RESULTS THAT MAY BE OBTAINED BY USING THE SOFTWARE OR CONTENTS OF THE BOOK. THE AUTHOR AND PUBLISHER HAVE USED THEIR BEST EFFORTS TO ENSURE THE ACCURACY AND FUNCTIONALITY OF THE TEXTUAL MATERIAL AND PROGRAMS CONTAINED HEREIN; WE, HOWEVER, MAKE NO WARRANTY OF ANY KIND, EXPRESS OR IMPLIED, REGARDING THE PERFORMANCE OF THESE PROGRAMS OR CONTENTS. THE SOFTWARE IS SOLD "AS IS" WITHOUT WARRANTY (EXCEPT FOR DEFECTIVE MATERIALS USED IN MANUFACTURING THE DISC OR DUE TO FAULTY WORKMANSHIP).

THE AUTHOR, THE PUBLISHER, DEVELOPERS OF THIRD PARTY SOFTWARE, AND ANYONE INVOLVED IN THE PRODUCTION AND MANUFACTURING OF THIS WORK SHALL NOT BE LIABLE FOR DAMAGES OF ANY KIND ARISING OUT OF THE USE OF (OR THE INABILITY TO USE) THE PROGRAMS, SOURCE CODE, OR TEXTUAL MATERIAL CONTAINED IN THIS PUBLICATION. THIS INCLUDES, BUT IS NOT LIMITED TO, LOSS OF REVENUE OR PROFIT, OR OTHER INCIDENTAL OR CONSEQUENTIAL DAMAGES ARISING OUT OF THE USE OF THE PRODUCT.

THE SOLE REMEDY IN THE EVENT OF A CLAIM OF ANY KIND IS EXPRESSLY LIMITED TO REPLACEMENT OF THE BOOK AND/OR CD-ROM, AND ONLY AT THE DISCRETION OF CRM.

THE USE OF "IMPLIED WARRANTY" AND CERTAIN "EXCLUSIONS" VARY FROM STATE TO STATE, AND MAY NOT APPLY TO THE PURCHASER OF THIS PRODUCT.

PREVENTATIVE PROGRAMMING TECHNIQUES

AVOID AND CORRECT COMMON MISTAKES

BRIAN HAWKINS

CHARLES
RIVER
MEDIA

CHARLES RIVER MEDIA, INC.
Hingham, Massachusetts

Copyright 2003 by CHARLES RIVER MEDIA, INC.
All rights reserved.

No part of this publication may be reproduced in any way, stored in a retrieval system of any type, or transmitted by any means or media, electronic or mechanical, including, but not limited to, photocopy, recording, or scanning, without *prior permission in writing* from the publisher.

Publisher: Jenifer Niles
Production: Publishers' Design and Production Services, Inc.
Cover Design: The Printed Image

CHARLES RIVER MEDIA, INC.
10 Downer Avenue
Hingham, Massachusetts 02043
781-740-0400
781-740-8816 (FAX)
info@charlesriver.com
www.charlesriver.com

This book is printed on acid-free paper.

Brian M. Hawkins. *Preventative Programming Techniques: Avoid and Correct Common Mistakes.*
ISBN: 1-58450-257-6

All brand names and product names mentioned in this book are trademarks or service marks of their respective companies. Any omission or misuse (of any kind) of service marks or trademarks should not be regarded as intent to infringe on the property of others. The publisher recognizes and respects all marks used by companies, manufacturers, and developers as a means to distinguish their products.

Library of Congress Cataloging-in-Publication Data

Hawkins, Brian M.
   Preventative programming techniques : avoid and correct common
mistakes / Brian M. Hawkins.
      p.   cm.
   ISBN 1-58450-257-6 (Paperback with CD-ROM : alk. paper)
   1. Computer programming.   I. Title.
   QA76.6.H388 2003
   005.1—dc21
                         2003001341

Printed in the United States of America
03 7 6 5 4 3 2 First Edition

CHARLES RIVER MEDIA titles are available for site license or bulk purchase by institutions, user groups, corporations, etc. For additional information, please contact the Special Sales Department at 781-740-0400.

Requests for replacement of a defective CD-ROM must be accompanied by the original disc, your mailing address, telephone number, date of purchase and purchase price. Please state the nature of the problem, and send the information to CHARLES RIVER MEDIA, INC., 10 Downer Avenue, Hingham, Massachusetts 02043. CRM's sole obligation to the purchaser is to replace the disc, based on defective materials or faulty workmanship, but not on the operation or functionality of the product.

This book is dedicated to my family, friends,
and two Keeshonds Austin and Libby.

Contents

Acknowledgments

First and foremost, I would like to thank my wife Debra for her patience and understanding of my crazy schedule and long work hours that allowed me to write this book. I would also like to thank Kim Lay, my best friend and fellow programmer, for her support and feedback on the concepts discussed in this book.

Many thanks go to Jenifer Niles and the rest of the production team at Charles River Media for the opportunity to make this book a reality. Without their patience and assistance, this book would have been much more difficult to finish.

This book would not have been possible without my experiences working at Justsystem Pittsburgh Research Center, Activision, Seven Studios, and consulting for JAMDAT. In particular, I would like to thank Scott Fahlman for the opportunity to learn the importance of research while at Carnegie Mellon University and Justsystem Pittsburgh Research Center. Also, thanks to Dr. Ian Lane Davis for the opportunity to begin working on the development side of software production.

Thanks to the many coworkers and friends with whom I have spent long hours discussing software development, including Chris Buchanan, Michael Douglas, Paul Gleichauf, Paul Haban, John Hancock, Ivanna Kartarahardja, Nick Kramer, Rita Lui, Donovan Mandap, Mike Mant, Steve Mariotti, John Miller, Gordon Moyes, Hamilton Slye, Rahul Sukthankar, and Andy Witkin. If I have left anyone out, please forgive me and know that I appreciate your contributions as well. In addition, thanks to the many other writers from whom I have learned much.

About the Author

Brian Hawkins graduated with a bachelor's degree in Mathematics and Computer Science from Carnegie Mellon University and immersed himself in computer graphics research at Justsystem Pittsburgh Research Center. After two years, his desire to hold a finished product in hand led him across the United States to join Activision in Los Angeles, where he worked as the game core and user interface lead on *Star Trek: Armada*. In addition, he contributed to *Civilization: Call To Power* and *Call To Power 2*.

The next step for him was to take a position as lead programmer at Seven Studios to help create *Defender*. Brian is now offering consulting services from his own company, Soma, Inc., and has recently worked on several mobile phone games and applications for JAMDAT Mobile.

Introduction

WHAT IS A PROGRAMMER ILLNESS?

Repeatedly I see the same mistakes being made during software development. These errors hide behind slightly different masks, but at their heart, many of them are the same. I have made many of these mistakes myself, not realizing that I was repeating the same errors until well after they had been made . Even one of these mistakes can have far-reaching consequences for a software development project; making them over and over is guaranteed to cause loss of time or money.

Many projects I have worked on started out with a goal to avoid past problems. Unfortunately, the projects often fell into the familiar programming traps with all the corresponding complications associated with them. One of the major problem areas projects fall into is having to perform considerable optimization work toward the end of the project. Knowing that this was a major problem for our particular project, my team established a mandate to optimize this new project from the very start. Optimizations were made early and were considered to be of primary importance. There was also an emphasis on early completion of features using whatever means necessary.

However, these methods led to several of the classic mistakes made by programmers. By focusing on optimization alone, the code was made difficult to understand and hard to modify. As the project went on, changes became increasingly difficult to make, and fixes to existing code took longer. This led to cutting of features and simplification of the project's scope. A considerable amount of extra work was also accrued due to the distrust of the performance of externally generated code, such as the C++ Standard Template Library, and higher-level language features. This further reduced the amount of new features that could be implemented, and made standard features more onerous to implement. Topping this off were the other common mistakes not related to the goal of optimization, such as cut-and-paste coding used to implement features quickly in the short term. All of this meant a considerable amount of time, and hence money, were spent on a small feature set.

So, was this reduction in scope worth the performance gained from this approach? Unfortunately, the desired performance improvements were not realized, resulting in the hiring of a consultant to optimize the application near the end of development. Despite the intentions and efforts of the developers, loss of time and difficulties in development were not offset by the desired performance goals. Although the project was completed, more could have been achieved in the same time, or the project could have been finished earlier. Even a successful project leaves a lot of room for improvement, and there is hope that many of the team members took away lessons from the failures in the project as well as the successes.

Seeing these same errors again and again is similar to how a doctor diagnoses the same illness over and over. Once an illness is diagnosed, however, methods for detecting, preventing, and curing the illness are developed. Similarly, programmers need to develop techniques for detecting, preventing, and curing common programming errors. This similarity inspired my ideas to use the illness metaphor for describing programming errors. This metaphor serves several purposes, most important as an aid to remembering the importance of each of these mistakes. Comparing programmer mistakes to a more common problem of human illness also allows each mistake to be discussed in a structured manner that aids in reference and recall. Programmer illnesses, just as regular human disease, can also be communicable. A new programmer is very likely to pick up these mistakes from his mentor. A programmer working on code filled with the little bugs will be less careful and might, through lack of caring or out of necessity, introduce similar bugs. These problems must be prevented, or caught and cured early to reduce their impact on your project.

On a slightly less serious note, some of the terminology already in use is reminiscent of the fancy terminology used by doctors. For example, Premature Optimization and NIH Syndrome sound like some strange disease or mental condition. In a way, these actually do describe a mental condition present in many programmers. To counteract this problem, you must be disciplined in your approach to design and coding. This importance of discipline is evident in the recent success of *Extreme Programming* and other agile methodologies. These new methodologies suggest an approach that requires fewer but more disciplined programmers to accomplish the same tasks that a larger group of average programmers would take the same amount of time to do. In addition, nobody wants to catch a bug, be it in code or in our blood.

WHO SHOULD READ THIS BOOK?

This book is intended to be useful for both novice and experienced programmers. Knowledge of basic programming concepts in at least one language is assumed. Fancy jargon is avoided when possible, and most of the information is self-contained. This book is intended to be practical, and therefore focuses on ways that you can reduce these problems in your everyday work. Before you dismiss these illnesses as common knowledge, take a closer look at code you are actually working with and see how often these mistakes are made. Many of you will be surprised at how often this common knowledge is not applied.

What might also be surprising to you is that the majority of the principles and practices espoused in this book are language independent. I have done practical work in C, C++, and Java, as well as learning a number of other languages, and have seen most of the illnesses present in all of them. Some languages can make it easier to prevent and cure these illnesses, but they do not provide a guarantee. For example, inheritance is a useful feature that can help reduce cut-and-paste, but there is nothing to prevent the programmer from ignoring it and simply using cut-and-paste anyway. In the end, the programmer is responsible for being disciplined and using the tools and language features at his disposal to prevent these illnesses from taking hold. No matter what language you are working in, this book is for you.

HOW TO USE THIS BOOK

It is strongly recommended that everyone read about the major illnesses. For the novice programmer, it is also a good idea to read this book from front to back. You will be encountering many of these decisions and problems in the years to come. For the experienced programmer, some of the material might be familiar to you, so feel free to skim it. However, try not to dismiss any of the sections outright, as there are many useful tips included that you might not have thought of for improving your coding. Some of the included anecdotes are also amusing, and would you want to miss those?

There are two main parts to the book: major illnesses and minor illnesses. The major illnesses are three of the most common problems in programming. Their effect is felt everywhere, and can cause very serious problems during development. Each major illness is discussed in detail and is likely to be the root cause of one or more of the minor illnesses. The minor illnesses are problems that are smaller in scope, but nonetheless still common and problematic for developers. Less common than the major illnesses, they can be equally destructive if they are not caught and handled early. Here is a short description of each of the illnesses:

Premature Optimization: Optimizing too early in development has many disadvantages, and can make development unmanageable toward the end of the project. Yet, this is one of the most common illnesses that plague programmers and projects.

CAP Epidemic: Duplicating editable information quickly leads to arduous maintenance tasks and nightmares when trying to make changes, but too often cut-and-paste is seen as a quicker and easier solution. Along with premature optimization, these make up the majority of programming errors I have seen.

NIH Syndrome: Fear is a powerful thing, and programmers fear any code they have not written. While this fear might occasionally be justified, we often throw the baby out with the bath water in our automatic reaction to using other programmer's code.

Complexification: Almost all programmers love a challenge, and coming up with a complex solution to a simple problem is such a challenge. The problem is that a much simpler solution will often do just as well.

Over Simplification: Those who have been bitten by Complexification often overcompensate by trying to remove all complexity, thereby making things too simple. This usually results in shifting complexity elsewhere, and often making the new code more complex than necessary.

Docuphobia: If we wanted to be writers, why did we major in computer science or mathematics? Unfortunately, when we do not document properly, others cannot use our code, and we might even forget what we did over time.

i: i? Are you having trouble figuring out what that means? Well, the same problem occurs when trying to read code with names like that. One might call it self-obfuscating code.

Hardcode: Can you afford to hire a programmer to sit with every other member of the design team anytime a change needs to be made? If not, then all those hard-coded values are going to cause you some problems.

Brittle Bones: Ever try to build a house on top of a swamp? It is similar to trying to build an application on top of buggy libraries and a brittle framework.

Requirement Deficiency: Do you want to reach the end of a project only to find that the customer wants more? To avoid this, make sure that all the requirements are laid out, even the ones that customers often forget to mention.

Myopia: Why solve today what you can put off until tomorrow? The problem is that you often have to do work to put the problem off when you could have solved it right the first time.

Each illness follows the same structured layout to make referencing easy. An illness is broken down into Description, Symptoms, Prevention, Cure, Related Illnesses, First Aid Kit, and Summary:

Description: Short description of the illness and its effects.

Symptoms: Details on how to spot this illness.

Prevention: Techniques to prevent this illness from causing errors.

Cure: Techniques to recover from problems caused by this illness.

Related Illnesses: Other illnesses that occur with or because of this illness.

First Aid Kit: Tools to help in the prevention and curing of the illness.

Summary: Summary of the illness and how to deal with it.

You should read the entire illness on the first pass, and then skip to the specific section later to aid in recalling particular information.

NO SILVER BULLET

There is no single development, in either technology or management technique, which by itself promises even one order-of-magnitude improvement within a decade in productivity, in reliability, in simplicity. [Brooks95]

No single technique can be applied to all situations in software development. Therefore, it is important to keep in mind that the rules and suggestions must always be considered in the current context of your project. The majority of the advice given in this book is meant to apply to most situations that you will encounter during development, but do not be afraid to ignore them when they do not apply. Just be sure you think carefully about your reasons for using a different approach. One useful method is to have a colleague play a devil's advocate whom you must convince that your solution is the correct one. Try to choose someone you feel would actually disagree with you at first; that way, you are not getting false reassurance from a fellow sympathizer.

The methods and techniques suggested in this book come from personal experience, correspondence with colleagues, and written material from books, magazines, and the Internet. Every reasonable effort has been made to provide accurate and useful advice, but as with any large endeavor, errors can creep in. Please feel free to e-mail me at *prevention@somaconsulting.com* with any corrections, insights, or opinions that you might have regarding this book.

Major Illnesses

The three major illnesses are Premature Optimization, CAP (Cut-And-Paste) Epidemic, and NIH (Not-Invented-Here) Syndrome. These represent the most commonly repeated mistakes that programmers from all skill levels are prone to make. They are also responsible for major losses in development time and money. While many agree that these are problem areas, they are still present throughout the industry due to a lack of knowledge about their full extent. The following chapters are intended to fill in much of this missing information and provide guidelines to help spot, prevent, cure, and otherwise alleviate these problems.

1 Premature Optimization

DESCRIPTION

Premature optimization encompasses any optimization that is made before it is required. Even though most programmers have been told not to perform premature optimization, very few fully understand the meaning or implications of these words. The occurrences of premature optimizations span a wide range, from low-level design to source code. The reasons why these optimizations are performed too early also span a great range.

This issue is often a very controversial one, but more often programmers err on the side of premature optimization. We therefore concentrate on the techniques that allow a programmer to avoid making premature optimizations, but this does not eliminate the need for optimizations altogether. The key word is *premature*, avoiding optimizations before they are necessary. As was stressed in the Introduction, there are no silver bullets. Keep in mind your project's particular performance requirements, or budget, and realize that there are some cases where design decisions in particular must be made at an earlier stage in the project. The emphasis is on ensuring that the optimizations are required at that stage, in which case they would not be premature. With this in mind, let us look at some of the symptoms that indicate that optimizations are being made prematurely.

SYMPTOMS

Premature optimization pervades computer science like a plague. You usually do not have to look very far to find an example of this problem. One of the most important steps in preventing premature optimization is to learn when it is being done. In this section, we will examine the varied symptoms that point to this illness.

First, the more general indicators that apply under all circumstances are presented. Then, several primary instances when premature optimizations can occur will be looked at. Finally, we look at reasons for optimization that are often false and can lead to premature optimization. Only when you know how to diagnose the problem can you move on to curing current problems and preventing future ones.

WHY?

So, if premature optimization has so many bad consequences, why do people still make this same mistake repeatedly? There are a wide variety of reasons that people give for performing optimization too early, and it is very important to know these reasons and why they are flawed. This knowledge will protect you from following these lines of reasoning and allow you to understand and discuss with other programmers why they might want to consider not performing an early optimization. As we explore the various symptoms of premature optimization, some of the major reasons for premature optimization will also be explored.

General Symptoms

There are several general indicators that premature optimization is occurring. Because of their broad nature, however, many of these signs only indicate that more investigation is necessary. Watch for these symptoms and, when they appear, delve further into their cause.

Readability

One of the most common indicators of premature optimization is poor code readability. The majority of optimizations are more complicated solutions and therefore harder to read and understand than a simpler but less optimal solution. This does not mean that all difficult to read code has been optimized. An unfortunately all too common reason for poor readability is simply poor code. It is necessary to examine the purpose of the code to determine if it is performing useful work that could be an optimization, or if it is simply hiding a much less complicated algorithm in highly obfuscated code.

The poor readability introduced by premature optimizations makes the code more difficult to understand and modify. Since every extra day a programmer spends modifying code costs money, you must balance the benefits of the performance gain over the loss in development time and increased cost of programming labor. It must be emphasized that there are few optimizations that do not hurt readability or robustness, so you should assume that you will be making this trade-off when making a decision. It should be obvious from the link between optimization and longer development time why it is best to wait until the latter stages of development to perform optimizations.

Take for example the following sorting algorithm, called selection sort, written in C++ and included in Source/Examples/Chapter1/readability.cpp on the companion CD-ROM:

```cpp
void selection_sort(int *io_array,
                    unsigned int i_size)
{
    for(unsigned int l_indexToSwap = 0;
        l_indexToSwap < i_size; ++l_indexToSwap) {

        unsigned int l_indexOfMinimumValue =
            l_indexToSwap;

        for(unsigned int l_indexToTest =
            l_indexToSwap + 1;
            l_indexToTest < i_size;
            ++l_indexToTest) {

            if(io_array[l_indexToTest] <
               io_array[l_indexOfMinimumValue]) {
                l_indexOfMinimumValue =
                    l_indexToTest;
            }
        }

        int l_minimumValue =
            io_array[l_indexOfMinimumValue];
        io_array[l_indexOfMinimumValue] =
            io_array[l_indexToSwap];
        io_array[l_indexToSwap] = l_minimumValue;
    }
}
```

This straightforward algorithm is easy to read and understand. However, a much more efficient sorting algorithm is the heap sort algorithm, also included in Source/Examples/Chapter1/readability.cpp on the companion CD-ROM:

```
void sift_down(int *io_array, unsigned int i_size,
               int i_value, unsigned int i_index1,
               unsigned int i_index2)
{
    while(i_index2 <= i_size - 1) {
        if((i_index2 < i_size - 1) &&
           (io_array[i_index2] <
            io_array[i_index2 + 1])) {
            ++i_index2;
        }
        if(i_value < io_array[i_index2]) {
            io_array[i_index1] = io_array[i_index2];
            i_index1 = i_index2;
            i_index2 *= 2;
        } else {
            break;
        }
    }

    io_array[i_index1] = i_value;
}

void heap_sort(int *io_array, unsigned int i_size)
{
    if(i_size < 2) {
        return;
    }

    for(unsigned int l_hire = i_size / 2;
        l_hire > 0; --l_hire) {
        sift_down(io_array, i_size,
                  io_array[l_hire - 1],
                  l_hire - 1,
                  (2 * l_hire) - 1);
    }

    for(unsigned int l_retire = i_size - 1;
        l_retire > 1; --l_retire) {
        int l_value = io_array[l_retire];
        io_array[l_retire] = io_array[0];
        sift_down(io_array, l_retire,
                  l_value, 0, 1);
    }
```

```
    int l_swap = io_array[1];
     io_array[1] = io_array[0];
     io_array[0] = l_swap;
}
```

While still not the most difficult to understand code, it will certainly take more time to parse than the simpler selection sort algorithm. This decrease in readability is only desirable if the performance gained is necessary.

One parting note, which we will discuss in more detail when we talk about NIH (Not-Invented-Here) Syndrome, is to avoid writing your own sorting algorithms in most cases. The sorting algorithms presented here were simply for example purposes; there is usually no reason to rewrite work that is likely to be freely available for almost every language.

Dependency

Optimization must sometimes cross encapsulation boundaries, leading to greater dependency between classes or modules that can reduce the robustness of the code. These dependencies make changes more difficult by causing each change to affect more of the code base than would otherwise be necessary. In some languages, such as C++, these dependencies might also cause other unwanted problems such as longer compile times. As with readability, code with too many dependencies does not indicate optimized code, but can also result from poor coding practices.

Look for two or more units, such as classes or modules, that are overly dependent on each other. Does changing one of the code units require a change to the other code unit? If so, chances are that these units are tightly coupled, possibly for the purpose of optimization. There are several things to consider at this point. First, should these units actually be separate, or can they be incorporated into one cohesive unit? Depending on the design, having them separate might have caused encapsulated data to become exposed. This is one of the least desirable results of such a dependency and should be the first to be remedied as we discuss in the "Cures" section. Second, if upon consideration it makes sense to have them as two separate units, does the dependency serve a useful purpose? This is where a mistake is often made; using an optimization as a justification is not the right decision. Very rarely does the performance increase outweigh the increase in dependency and the resulting brittleness of the code.

To help illustrate such dependencies, let us provide a simple example using pseudo-code. The goal of the code is to call a launch function that will launch a missile at a target when the target is in range. This is a common occurrence in many games, including jet fighter simulations. The launch function takes a unit vector pointing in the direction the missile should launch, which would be defined something like this:

```
launch(vector unit_direction);
```

Now we can create the function that checks the range and calls launch when we are in range. This function requires the origin of the missile launcher, the target destination, and the range of the missile. The range check can be done by taking the dot product of the vector from the origin to the destination with itself, because the dot product of a vector with itself is the vector's length squared. The range must also be squared to make the comparison valid. If the launch is a go, the vector from the origin to the destination must be normalized, or made into a unit vector, before it is passed to the launch function. This is accomplished by dividing the vector by its length, or the square root of the dot product of the vector with itself. This results in the following function:

```
launch_when_in_range(vector origin,
                     vector destination,
                     real range)
{
    if(dot_product(destination - origin,
                   destination - origin)
       is less than (range ^ 2)) {

        launch(normalize(destination - origin));
    }
}
```

Because we have not optimized the function yet, it is easy to see that we can reuse the range check by placing it in a separate function:

```
bool is_in_range(vector origin,
                 vector destination,
                 real range)
{
    return whether
        dot_product(destination - origin,
                    destination - origin)
        is less than (range ^ 2);
}
```

```
void launch_when_in_range(vector origin,
                          vector destination,
                          real range)
{
    if(is_in_range(origin, destination, range)) {
        launch(normalize(destination - origin));
    }
}
```

Perhaps it turns out that this function is called often, so when we get to the optimization phase we decide to improve its performance. A simple observation indicates that we are computing the same values multiple times, so to improve the performance, we place them in temporary variables that can be reused throughout the function:

```
void launch_when_in_range(vector origin,
                          vector destination,
                          real range)
{
    vector origin_to_destination =
        destination - origin;
    real origin_to_destination_length_squared =
        dot_product(origin_to_destination,
                    origin_to_destination);

    if(origin_to_destination_length_squared
       is less than (range ^ 2)) {

        launch(origin_to_destination / square_root(
            origin_to_destination_length_squared));
    }
}
```

This function has a much tighter coupling between its various parts, making it difficult to separate any part out for reuse. It is also more difficult to read than the two separate functions that we were able to separate the original function into. If this version had been written early in development, it is likely that it would not have been refactored into the separate functions. This would have led to a duplication of the range check elsewhere, leading to code that is more difficult to maintain and modify. This example illustrates dependencies on a functional level, but they can also occur at other levels such as class or module interdependencies. Look carefully at all interdependencies to see if they represent premature optimization.

Constraints

Another indicator of premature optimization is overly constrained data input, leading to an inflexibility that will not allow changes later in development. If a more general solution is evident, then it is likely that the code was written to be optimized rather than robust. As with other signs of optimization, it is possible to choose an algorithm that exhibits too many constraints without a performance gain. However, this is generally less common with this particular symptom. Worse than simply constraining the inputs from being accepted, the usual result is an algorithm that accepts the inputs but performs worse than the more generic algorithm that does not constrain the data.

Let us illustrate this with a simple example involving hash tables. A *hash table* is an indexed array where the index is determined by applying a hash algorithm to a key from a large range to obtain an index within the range of the array. The key focus of this example will be the hash algorithm, but to see its effect on performance, you must understand the other parts of the algorithm. One of the results of taking the keys from a larger range and converting them to the smaller range of the indices is the possibility of a conflict occurring. There are several methods of resolving these conflicts. For this example, we will assume that the data set is smaller than the array, so we can simply perform a linear search from the initial hash position until we find a valid spot. For inserting a valid spot would be an empty spot, and for searching a valid spot would contain an exact match for the search key. Here are the insert and search functions:

```
insert(string key, data)
{
    base_index = index = hash(key);
    while array[index] is not empty
    {
        index = (index + 1) modulus array_size;

        // Assume data fits in array, if the index
        // wraps around to the original index the
        // array is filled and our assumption has
        // been violated.
        assert(index is not base_index);
    }
    array[index] = data;
}


search(string key)
```

```
    {
        base_index = index = hash(key)
        while array[index].key does not match key
        {
            if array[index] is empty
                return item_not_found;

            index = (index + 1) modulus array_size;

            if index matches base_index
                return item_not_found;
        }
        return array[index];
    }
```

Notice that the more keys that hash to the same location, the longer insertion and searching will take for those keys. Thus, it is important to ensure that the keys in the input data set hash to different locations as often as possible. Take the following set of keys:

```
Ada Hawkins
Brian Hawkins
Chester Hawkins
Debra Hawkins
Kathleen Hawkins
Loreena Hawkins
Robert Hawkins
Trip Hawkins
```

Given this set of keys, you might be tempted to write a hash algorithm that performs fast with the given input:

```
hash(string key)
{
    return key[0] modulus array_size;
}
```

This hash will perform well with a given set of keys, which follow the constraint that most or all of them start with different letters. Now imagine the set of keys changed to the following:

```
Hawkins, Ada
Hawkins, Brian
```

```
Hawkins, Chester
Hawkins, Debra
Hawkins, Kathleen
Hawkins, Loreena
Hawkins, Robert
Hawkins, Trip
```

Our hash algorithm will now generate the same index for every key, eliminating the advantage of using a hash table and reverting it to a simple linear search. The worse part about this type of premature optimization is the possibility that external data changed after the application is complete can radically influence the performance of the application. If instead the following hash algorithm [Sedgewick90] were used, the performance would not be optimal for the initial set of keys:

```
integer hash(string key, integer array_size)
{
    hash = 0;
    for each character in key
        hash = ((64 * hash) +
            character) modulus array_size;
    return hash;
}
```

However, any change to the keys would result in little change in performance, making the algorithm's performance more predictable and higher on average than many of the other algorithms. If the performance improvements do become necessary toward the end of development during the optimization stage, be sure to provide a warning or error for data sets than cause considerable performance loss. For example, the insert function might be rewritten like this:

```
insert(string key, data)
{
    count = 0;
    base_index = index = hash(key, array_size);
    while array[index] is not empty
    {
        index = (index + 1) modulus array_size;

        count = count + 1;
        if count is too large
            warn hash_algorithm_performance_reduced;
```

```
        // Assume data fits in array, if the index
        // wraps around to the original index the
        // array is filled and our assumption has
        // been violated.
        assert(index is not base_index);
    }
    array[index] = data;
}
```

Performance Tradeoffs

At this point you might be thinking, "Well, sure these optimizations lead to code that is harder to read and modify, but aren't the performance improvements worth the tradeoff?" The answer is "not usually." The fastest way to understand why is to profile the entire application, and find out what percentage of time is spent in each of the different algorithms and modules. On average, 80 percent of the time should be spent in 20 percent of the code, and conversely, 80 percent of the code is only called 20 percent of the time. For simplicity's sake, imagine that every optimization took the same amount of development time, and for every optimization, you lose one feature. You should already notice that you would probably not want to make every optimization, but concentrate on the ones that affect the 20 percent of the code that is running 80 percent of the time. This is even before we consider when the optimization is made. Suppose also that for every month after an optimization is made, but before you ship, you lose one additional feature because of the difficulties caused by poor readability and code fragility. Obviously, you would want to wait until later in the project and make only those optimizations that are necessary to meet your performance goals. While we have simplified our description to make it easier to visualize, this is exactly the type of tradeoff that must be considered on real projects.

What all this really means is that you should consider the real performance benefits of an optimization against all the disadvantages of that same optimization before deciding to go ahead with it. Another of the considerations, and a reason to watch for optimizations that are made too early, is whether the optimization will even be there when the project is finished. The earlier you make an optimization, the more likely it will not be there by the end. A good way to see exactly how many real optimizations you are making is to keep a record of each optimization made. Then, at the end of the project, review the code to see how many of them are still in place. If substantial amounts of these optimizations are no longer there, then you need to adjust your optimization policies for the next project.

WHY LATER IS TOO LATE

So, you have found this situation where it would be much more difficult to perform the optimization later, and you are sure that this is not just habit speaking. These situations do occur, but take a very careful look at the reasoning before going ahead with the optimization. Valid scenarios for early optimization do not come along often, so be sure there is plenty of evidence backing this decision or you will regret it later. The more likely reason for difficulty in making optimizations later is poor programming practices. *Spaghetti code*, or code where every part is dependent on every other part, is one of the most common problems that makes optimizations difficult. However, there are plenty of other reasons.

The other reason to be suspicious of this claim stems from the turbulent nature of software development. Change is common during programming, and even if the assumption that a portion of the code will be difficult to optimize later is correct, that code might be called only occasionally or not at all by the end of the project. If this is the case and the optimization was made anyway, programmer time was lost on a needless task that could have been avoided. It is often better to trade the risk of having to do the optimization later for the guaranteed loss of time and code complexity of making the optimization up front.

False Optimizations: The Optimization that Wasn't

Programmers are almost always very poor judges of what optimizations will actually make the application run substantially faster. This has caught most programmers repeatedly. However, there is something even worse than wasting time on small optimizations; some supposed optimizations actually reduce the performance of the program. If you profile and the application is running slowly during an algorithm that is theoretically fast, then it is time to look at some practical issues that can cause this. A very important warning, however, is to save this process for as near the end of development as possible in order to maintain readability throughout development. In addition, small changes can easily affect these performance times, so early work might be wasted.

Among the more common reasons for false optimizations are either cache misses or page faults. The *cache* is a small piece of very fast memory located directly on the main processor (Figure 1.1). A cache miss occurs when the algorithm either produces a machine instruction set that cannot be stored in the processor in its entirety, or it accesses data that cannot all be stored in the cache. These cache misses can take longer than the extra processing required for a theoretically slower algorithm, resulting in better real performance from the slower algorithm. It is difficult to predict when this will occur, and small changes can easily result in very different

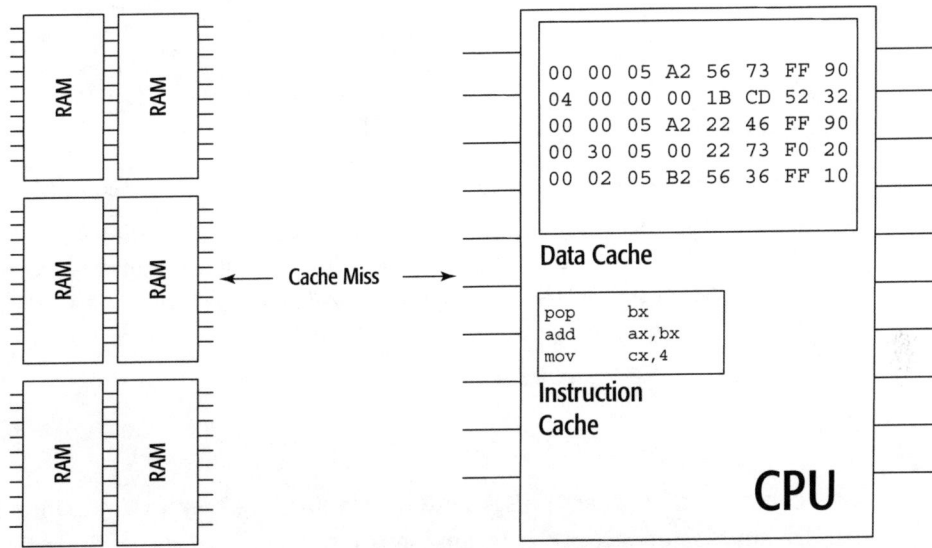

FIGURE 1.1 Data and instruction caches are located on the main processor, but must access memory when a cache miss occurs.

results. This makes it important to save any attempt at this type of optimization until the very end of development.

Page faults result from a similar set of circumstances. With most computers using virtual memory these days, data is often paged to disk when it does not all fit in physical memory at once. What this means is that an algorithm that uses a substantially larger amount of memory for performance improvements could be hit by slow I/O times (Figure 1.2). Again, this results in a theoretically faster algorithm achieving much slower performance in real applications.

Another programming technology, which continues to gain in popularity, that is the cause of many false optimizations is *threading*. Threading allows a process to execute multiple tasks at once, but complications arise when these tasks need to share data. This data sharing is one of the main differences between multitasking, running multiple processes simultaneously, and multithreading. Among other reasons, the number of multiple processor systems is continuing to increase, which

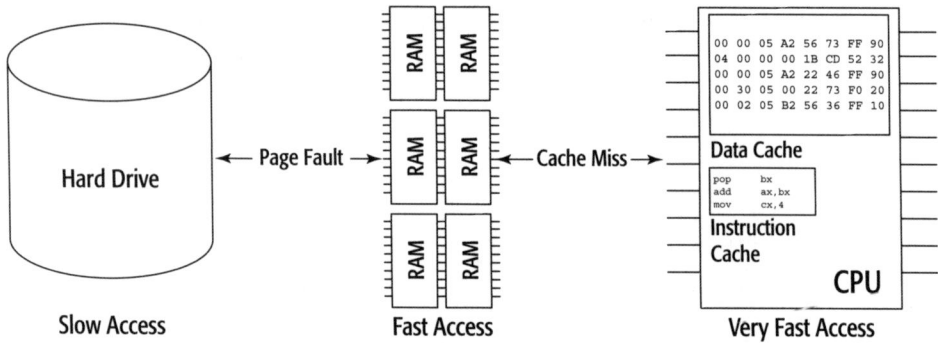

FIGURE 1.2   Data directly on the main processor is accessed the fastest, while data accessed from memory because of a cache miss is slightly slower, and data accessed from a hard disk or similar storage medium because of a page fault is substantially slower.

will continue to increase the use of multithreading. It is therefore very important to understand the impact of this technology on performance and optimization.

Multithreaded applications are particularly sensitive to timing issues and other hard to predict interactions between the various threads, making understanding the behavior of the application exponentially more difficult to understand and predict. Because of this, it is even more important to avoid complex algorithms and methods in places where threads interact. Look for these excessive data sharing and method calls between threads to find these problem areas. In addition, look for common single-threaded optimizations that are made without much thought, as these can often have the opposite result on multithreaded applications. A common example of this is *copy-on-write* (COW) optimizations for the standard template library string class that is popular with many implementers (Figure 1.3).

Several more false optimizations occur because of different system architectures and modern optimized compilers. These should only be dealt with toward the end of a project, but it is important to recognize areas where they might occur in order to take preventative measures that will assist these later optimizations. We will go into more detail about how to prepare your code when we talk about prevention.

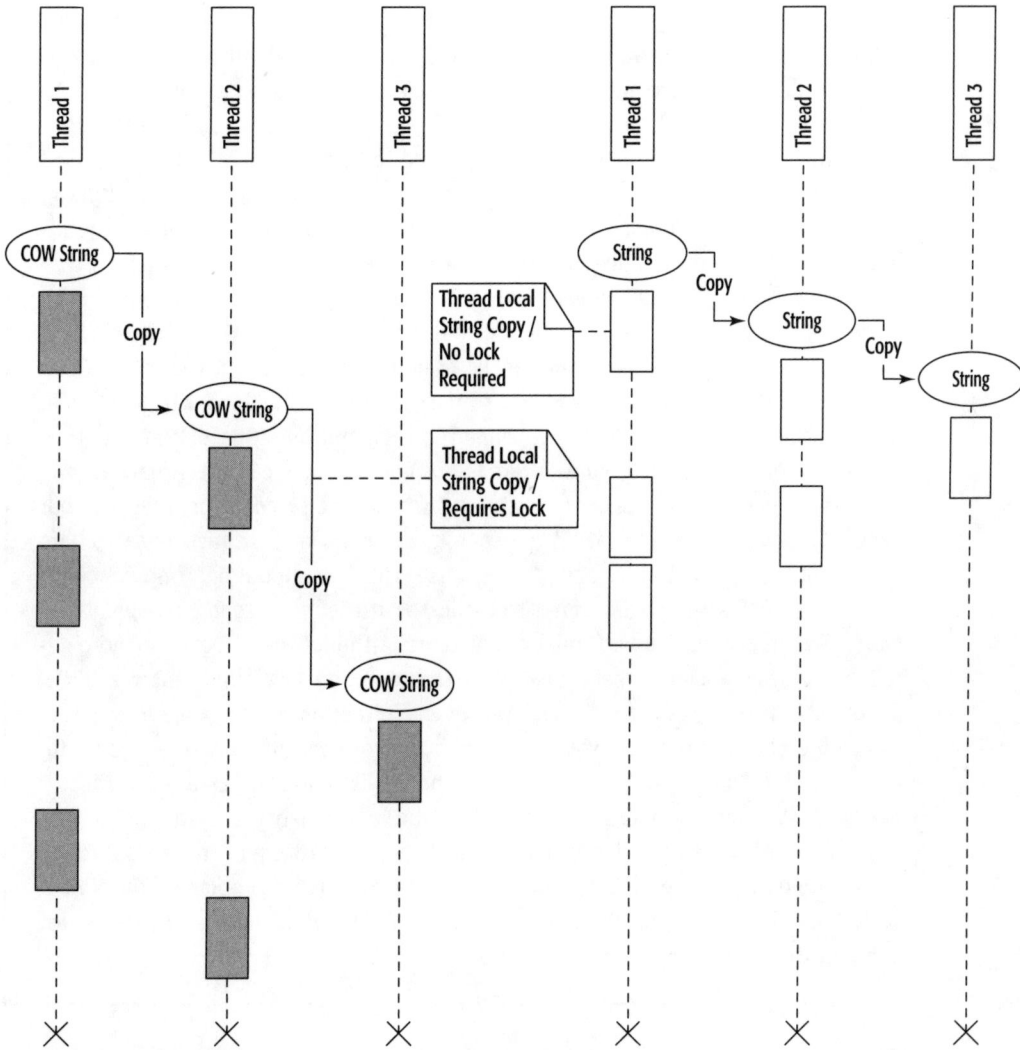

FIGURE 1.3 The left side shows a sample sequence using a copy-on-write (COW) string, while the right side shows the same sample sequence using a string that always performs a copy. Because the COW string shares data between threads, a lock must be performed for every copy, including local copies. This results in a longer running time for the overall application using the COW string.

WHY THAT HAS TO BE SLOW

One of the most common fallacies that programmers routinely entertain is that we know what will be optimal. Repeatedly, the same guesses about what is slow and how it could be fixed are made. Very few of these guesses have been correct. When it comes to the topic of optimization, intuition usually fails. Despite the fact that we are working with machines that are supposed to be deterministic, the best word to describe them is *quirky*. Computers often behave in ways we don't expect, and the large software programs and complex operating systems of the modern age interact to form emergent behavior that is often unexpected. It does not help that hardware designers use all types of unusual tricks to improve performance. Even worse are the hardware design decisions that are made to improve benchmark performance, which then become hindrances to more practical applications.

This belief will also leads us to one of the other major illnesses, NIH Syndrome, when the programmer falls into two incorrect assumptions. One of the beliefs is that only you can possibly know what is optimal for your current situation, and there is no possibility that code written by someone else could possibly fit your needs. The other belief is that all the code must be optimal. In fact, once you are convinced that the second assumption is false, the first should be of no concern.

One particular project decided to avoid the use of the *Standard Template Library* (STL) in part due to performance concerns. Although we will go into more detail about the problems that arose from this decision later in Chapter 2, "NIH Syndrome," it is relevant to mention its adverse effect on performance in this section. By eschewing the use of the STL, no standard containers were available for use in storing collections of objects. An in-house list class was created to fill this need, but it supported only a small subset of the functionality available in the STL. Because of this missing support, most uses of the list class simply iterated through the entire list. In the end, an attempt to avoid performance problems from a well-tested standard library led to even more performance problems along with a host of other disadvantages.

Caching

One common optimization that is often misunderstood, and consequently overused is caching of values that can be recomputed from already stored data. Do not fall for the temptation of caching on the first pass, but wait for the optimization phase of the project to determine which values would benefit from caching. In order to appreciate why this is important, let us examine an example of value caching and the consequences of this technique. Code for this example is also included in Source/Examples/Chapter1/cache_opt.h on the companion CD-ROM.

ON THE CD

Starting with a simple rectangle C++ class:

```cpp
class rectangle
{
public:

    rectangle(double i_length, double i_width) :
        m_length(i_length),
        m_width(i_width)
    {
    }

    void set_length(double i_length)
    { m_length = i_length; }

    void set_width(double i_width)
    { m_width = i_width; }

    double get_length() { return(m_length); }

    double get_width() { return(m_width); }

private:

    double m_length, m_width;

};
```

We then want to add a member function to obtain the area of this rectangle:

```cpp
class rectangle
{
public:

    // ...

    double get_area()
    {
        return(m_length * m_width);
    }

    // ...

};
```

Now, to avoid making the calculation for area every time the area is requested, we can store the area in the class. There are two common ways this can be done, the first of which involves computing the area any time the length or width of the rectangle changes:

```cpp
class rectangle
{
public:

    rectangle(double i_length, double i_width) :
        m_length(i_length),
        m_width(i_width)
    {
        update_area();
    }

    void set_length(double i_length)
    {
        m_length = i_length;
        update_area();
    }

    void set_width(double i_width)
    {
        m_width = i_width;
        update_area();
    }

    // ...

    double get_area() { return(m_area); }

private:

    void update_area()
    {
        m_area = m_length * m_width;
    }

    double m_area;

    // ...

};
```

Another familiar method is lazy evaluation, or waiting for the area to be requested, which can be done with some minor modifications to the previous class:

```cpp
class rectangle
{
public:

    rectangle(double i_length, double i_width) :
        m_length(i_length),
        m_width(i_width)
    {
        area_needs_update();
    }

    void set_length(double i_length)
    {
        m_length = i_length;
        area_needs_update();
    }

    void set_width(double i_width)
    {
        m_width = i_width;
        area_needs_update();
    }

    // ...

    double get_area()
    {
        if(m_area_needs_update) {
            update_area();
        }
        return(m_area);
    }

private:

    void area_needs_update()
    {
        m_area_needs_update = true;
    }

    void update_area()
```

```
    {
        m_area = m_length * m_width;
        m_area_needs_update = false;
    }

    bool m_area_needs_update;

    // ...

};
```

Lazy evaluation is not very effective in this case for a number of reasons, but it is shown because it is common and can often be useful in many other more complex cases. The primary reason for this ineffectiveness is the simplicity of the calculation versus the overhead in doing lazy evaluation. Trading off computation time in one form for another form that appears to be faster, but is not, is a common mistake.

The benefits of caching are obvious in this example; the get_area() function only need return a member value with no further calculations. What might be less obvious are the drawbacks to this method, which lead us to the assertion that it should only be used during the optimization stage. The most evident drawback is the added complexity to the rectangle class, making it less readable and therefore harder to modify. This alone should be argument enough for using the simpler form until near the end of development.

If that is not convincing enough, then consider the other consequences of caching data, primary among which is the extra memory requirements. The computed value must be stored, and if lazy evaluation is used, an additional flag must be stored to indicate when the stored value requires updating. This is particularly important on small systems that have memory limitations, but can even have an effect on larger systems due to a variety of possible slowdowns involving processor caches, memory transfers, boundary problems, and other unexpected problems.

While the problems with memory could lead to false optimizations, using caching can guarantee false optimization if the usage pattern of the data or class is not taken into account. To illustrate what this means, let us imagine using the rectangle class in a real-world application that is for editing the layout of rooms in a house. The user can change the size of any room interactively by dragging the sides or corners, and the other rooms adjust by shrinking or expanding. While this is happening, the set_length() and set_width() functions are being called multiple times for each user movement. Once the user is satisfied, he can print out a report with important information such as the area of each room. During this phase, set_area() is called once for each room. In this scenario, caching will cause extra work during the interactive phase and will only slightly speed up the report generation. Overall, this will lead to lower performance and an unsatisfied user. Consider the usage pattern before caching data even during the optimization phase, and always profile to determine the effects.

At this point, you might notice that lazy evaluation could help with this problem, since it would only introduce one extra instruction for each set called. As with all optimizations, this should still be avoided until the optimization stage. After all, those extra instructions are still not necessary in the case given, and the overhead of the extra `if` branch instruction can be large on many architectures. This does lead to an interesting point about developing general-purpose libraries, which is the one case that it might be useful to perform optimizations that are not tailored to a specific usage. There is no way to know the exact usage of the library in future applications, so optimizations must be carried out here as well. However, this should still be done near the end of the project, which in this case would be library deployment. Later, we will discuss how to approach this type of optimization in a slightly different manner than ordinary applications.

Since caching represents a very common symptom of premature optimization, it is important to always be on the lookout for values that can be computed instead of stored. A similar type of value that can appear in object-oriented language is the data member used as a temporary. This is done to eliminate the need to pass the value between functions within the class or through other classes. This should be avoided when possible, as it leads to many of the same problems as caching, which in essence it is.

WHY BAD HABITS ARE HARD TO BREAK

Back when memory was scarce and CPU speeds were measured in hertz instead of megahertz, it was necessary to ensure that every bit of code was efficient and small. Things have changed, but many of the lessons learned are still ingrained in the programmers and the teachings of the current day. They are like habits that we cannot seem to break, and these habits have become more than obsolete and are detrimental to modern software development. Even when we try to break free of these habits, we are often scared back when some young upstart programmer makes a mess of a project while trying to use fancy new techniques. Unfortunately, there are still a lot of programmers out there blindly following this technique or that technique rather than determining which one applies to the current situation.

Watch out for colleagues and mentors espousing the idea that this method or

(Continues)

(Continued)

that method will solve all the problems. Object-oriented programming will not make development simple. Functional programming will not make development simpler. Thinking is what makes development work, and considering all the factors in a situation is necessary to making things easier and faster. However, remember that mistakes are inevitable. The real task is to continue to improve development, and minimize the impact of mistakes and changes.

Speaking of mentors, they are very important to progressing in your abilities as a programmer. A good mentor can help you advance in both your abilities and your career at a rapid pace. You should choose your mentor wisely, looking for a senior programmer who is both a successful practical programmer and communicates well. One other important trait to look for in a mentor is a willingness to continue learning and discuss new ideas. A note to both mentor and student, always listen and learn from the other. Mentoring should be a two-way street, with both sides learning from each other.

Before you can help other programmers break their bad habits, you must first break yours. A useful start occurs just after you choose an algorithm for a programming task. Before you start coding, step back and consider your reasons for choosing a particular algorithm. The algorithm should meet the goals of the task while remaining simple to implement and modify. If you instead chose the algorithm with the highest performance, step back and take another look at the algorithms available. Is there one that is simpler to implement, or one that would make changes easier? Often, these will be the same algorithm, but either way you decide, always prefer the simpler choices to the algorithm with the highest performance until later during the optimization phase. Do not try to guess which algorithms should be optimized from the start; chances are you will be wrong and wasting your time.

Types of Optimization

Once beyond the general symptoms, it is important to understand the different varieties of optimizations. Any of these types can be performed before they are required, and learning to spot each type of optimization is essential to going from a general symptom to a determination that Premature Optimization is at work. Table 1.1 shows a comparison of the different optimization types. Let us look at each of these types in more detail.

TABLE 1.1 Comparison of Different Optimization Types

Type of Optimization	Description	Common Premature Optimization	Platform Specific	Language Specific
Low-level	Takes into account the assembly and machine code generated.	Yes	Yes	Yes
Algorithm	Choosing a more efficient algorithm, generally based on big O notation.	Yes	No	No
Memory	Deals with memory limitations of the target platform, including size, caching, and page faults.	No	Yes	No
Design	Changes that affect the overall applications, including the end user's experience, but that improve performance.	No	No	No
Source code	Minimizes the amount of text used to write the source code.	Yes	No	Yes

Low-Level

Low-level optimizations are usually localized and are language and system specific. Another important trait of this type of optimization is its susceptibility to small changes. Such things as cache size and memory alignment, as well as other esoteric hardware and operating system specific factors, influence low-level optimizations. These types of optimizations are the most likely to be unintuitive and fragile and should usually be saved until all other optimizations have been performed.

An example of such a low-level optimization is removal of the loop invariant. Take this small snippet of C/C++ code:

```
for(int index = 0; index < total; ++index)
{
    volume[index] =
        length * width * height[index];
}
```

Notice that length and width are not dependent on the loop variable. We can therefore remove the length and width multiplication from within the loop to avoid performing the computation on every iteration:

```
float area = length * width;
for(int index = 0; index < total; ++index)
{
    volume[index] = area * height[index];
}
```

This type of optimization gives large performance gains only if the code is called a noticeable percentage of the time. In addition, this type of optimization lends itself well to automation. Modern compilers already perform many low-level optimizations automatically, potentially making any work done in the source code unnecessary or even detrimental.

Algorithm

Algorithmic optimizations are the bread and butter of optimizations. They easily give you the most gain for your effort when it comes to performance improvements, and any programmer who is good at optimizations will find himself performing these types of optimizations first. Why waste time doing painful optimizations on some complex set of equations when you could greatly reduce or eliminate the calls to these equations by changing the higher-level algorithm used?

A simple illustration of an algorithmic optimization is a binary search versus a straightforward linear search. First, we look at the simple linear search that might be used initially:

```
find(search_criteria)
{
    current_item = item_at_head_of_list;
    while current_item is in list
    {
        if current_item matches search_criteria
            return current_item;
        current_item = item after current_item;
    }
    return item_not_found;
}
```

Now let us look at the higher performance binary search algorithm:

```
find(search_criteria)
{
```

```
left_index = 0;
right_index = number_of_items;
while left_index is not right_index
{
    middle_index = left_index + right_index / 2;
    middle_item = item[middle_index];
    if middle_item matches search_criteria
        return item[middle_index];
    if middle_item is less than search_criteria
        right_index = middle_index;
    if middle_item is greater than search_criteria
        left_index = middle_index;
}
return item_not_found;
}
```

Since many algorithms are language independent, this example is presented in pseudo-code to reinforce the cross-platform nature of this type of optimization. There are several other important points that stem from these examples, related to the symptoms discussed earlier. Let us take them one at a time.

First, the second algorithm is much longer and less readable than the first algorithm. In this case, it is only slightly longer, but with a large increase in the number of conditionals that make the logic harder to follow. Further optimizations to search algorithms can be even less readable, such as hash tables.

Next, the second algorithm adds a number of constraints to the item container that is being searched. The container in the linear search could be unsorted, while the binary search requires that the items in the container be sorted. Additionally, the items in the linear search need only be accessible sequentially, whereas the binary search requires random access through an index. Even if the current context of this algorithm meets these requirements, this limits the types of changes that can be made to the usage of the algorithm in the future.

Finally, we must consider the context of the algorithm's use. Notice that the binary search is longer, and therefore translates to more machine instructions than the linear search. If the number of items to search is small, this can lead to a false optimization that results in longer search times for the supposedly faster algorithm. For other algorithms, a similar phenomenon can occur with particular input sets.

WHY THAT NIFTY NEW ALGORITHM

Programmers must constantly solve difficult problems every day, so it is not surprising that the people most likely to take up the profession are the ones who enjoy finding solutions to complex problems. However, this can lead to a problem when faced with situations in which there is only a simple problem or no problem at all to solve. Faced with this, it is often tempting to invent problems that do not exist. This phenomenon is common and problematic enough to deserve its own minor illness, but it is introduced here because one of the common results is a desire to find a more optimal algorithm for code that is called less than 1 percent of the time. This brings along with it all the disadvantages of optimization with no noticeable performance gain.

Exacerbating this problem is the desire to try out a new algorithm or technique that the programmer has just heard about. This can have even worse effects than simply looking for the most optimal algorithm, because the programmer will often try to force an algorithm into a situation where it is not appropriate, thus resulting in a false optimization. Always consider carefully if the algorithm fits all the requirements of a given situation, especially future changes. As a programmer, you must always be learning new techniques and algorithms, but be particularly vigilant when you go to use an algorithm you have just learned.

One such nifty algorithm is *ternary string search* [Bentley98], which provides fast insertion and search times for looking up values based on a string key. With the proper data set, it also has a small memory footprint. Upon encountering this algorithm, one project decided to use it because they expected to need the performance benefits offered. Initially needing only insertion and search, a C++ implementation included in Source/Examples/Chapter1/nifty.h on the companion CD-ROM and similar to the following was used:

```
template <typename t_DataType> class t_StringMap
{
private:

    class t_MapNode
    {
    public:
        t_MapNode() : m_character(0),
            m_left(0), m_middle(0),
            m_right(0) {}

        char m_character;
```

ON THE CD

(Continues)

```
            t_MapNode *m_left;
            union {
                t_MapNode *m_middle;
                t_DataType *m_data;
            };
            t_MapNode *m_right;
        };

public:

        t_StringMap() : m_root(0) { }

        bool insert(const char *i_string,
                    t_DataType *i_data)
        {
            t_MapNode *l_node = 0;
            t_MapNode **l_nodePointer = &m_root;

            // Search through until we find an
            // empty spot.  The empty spot will
            // be stored in l_nodePointer so it
            // can be filled in the subsequent
            // section.
            while((l_node = *l_nodePointer) != 0)
            {
                if(*i_string < l_node->m_character)
                {
                    l_nodePointer =
                        &(l_node->m_left);
                } else
                if(*i_string > l_node->m_character)
                {
                    l_nodePointer =
                        &(l_node->m_right);
                } else {
                    // If we reach the end of the
                    // string, then this string
                    // is already in the list.
                    if(*i_string++ == 0) {
                        l_node->m_data = i_data;
                        return(true);
```

(Continues)

```
                                }

                        l_nodePointer =
                            &(l_node->m_middle);
                    }
                }

                // Insert the new characters,
                // start at the pointer we
                // found and countinue down
                // the middle.
                for(;;) {
                    *l_nodePointer = l_node =
                        new t_MapNode;
                    if(!l_node) {
                        return(false);
                    }
                    l_node->m_character = *i_string;

                    // If this is the end of the
                    // string, set the data and
                    // break out of our loop.
                    if(*i_string++ == 0) {
                        l_node->m_data = i_data;
                        break;
                    }
                    l_nodePointer = &(l_node->m_middle);
                }
                return(true);
            }

        t_DataType *find(const char *i_string) const
        {
            t_MapNode *l_node = m_root;

            // Continue until we run out of
            // nodes to search or we find
            // what we are looking for.
            while(l_node) {
                if(*i_string < l_node->m_character)
                {
                    l_node = l_node->m_left;
```

(*Continues*)

```
                        } else
                        if(*i_string > l_node->m_character)
                        {
                            l_node = l_node->m_right;
                        } else {
                            if(*i_string++ == 0) {
                                return(l_node->m_data);
                            }
                            l_node = l_node->m_middle;
                        }
                    }

                    // Match not found.
                    return(0);
                }

            private:

                t_MapNode *m_root;

            };
```

Already this is a little complicated to read and understand, but it did perform well and for a long time no changes were needed. However, later during development it became necessary to perform other operations on the key strings that required parsing the entire set. To do this required either the use of recursion or a stack, as this example function shows:

```
template <typename t_DataType> class t_StringMap
{

    // ...

    void print(t_MapNode *i_node = 0,
        char *i_string = 0,
        unsigned int i_index = 0)
    {
        char l_string[256];
        if(!i_node) {
            if(!m_root) {
                cout <<
```

(Continued)

```
                            "t_StringMap: ***  Empty  ***"
                                << endl;
                            return;
                        }
                        cout <<
                            "t_StringMap: --- Strings ---"
                            << endl;
                        l_string[0] = 0;
                        i_node = m_root;
                        i_string = l_string;
                    }

                    if(i_node->m_left) {
                        print(i_node->m_left,
                            i_string, i_index);
                    }
                    if(i_node->m_character) {
                        i_string[i_index] =
                            i_node->m_character;
                        i_string[++i_index] = 0;
                        print(i_node->m_middle,
                            i_string, i_index);
                        i_string[--i_index] = 0;
                    } else {
                        std::cout << i_string << std::endl;
                    }
                    if(i_node->m_right) {
                        print(i_node->m_right,
                            i_string, i_index);
                    }
                }

                // ...

            };
```

In this case, the desire to prematurely optimize made later changes more difficult. This used up precious programmer time that could have been spent on problems that were more important. The proper solution in this case would have been to use the containers in the standard template library, avoiding both Premature Optimization and NIH Syndrome, which we will discuss later.

Memory

On architectures with low amounts of memory, such as organizers and hand-held games, it might also be necessary to optimize the memory footprint of the application. Memory optimizations might also be necessary due to other hardware architecture quirks, such as alignment or transfer size requirements. As with other optimizations, these should not be performed until later in the development cycle if possible. However, there is a greater chance on small memory platforms that the application might exceed the memory requirements before the project is near completion. Just as you would optimize the largest CPU consumers when dealing with performance, address the largest memory users first until there is sufficient room to continue development. Do not overdo the amount of optimizations performed, save the remainder for later in development. Furthermore, if the memory improvements are purely for performance optimizations and not because the application is running out of memory, save them for near the end of development in the optimization phase.

While some memory optimizations can actually improve performance, most tend to involve a tradeoff between memory and performance. For example, take the decision between using bit-fields or booleans in C++ to represent boolean values in a class or structure. A bit-field uses one bit per value, usually with the minimum number of bits that can be used equal to the bit size of an integer. Thus, the number of bits wasted is only the integer bit size minus the total number of bits used. Booleans, on the other hand, usually use a full integer for each value, wasting the number of values multiplied by one minus the bit size of an integer. If there is more than one boolean value to represent, the bit-fields will use less memory. More values mean more memory saved. On most machines, setting and clearing bit-field flags and boolean flags would amount to the same performance, requiring one operation. However, on most architectures, testing a bit-field requires two operations, whereas testing a boolean requires only one operation (Figure 1.4, A and B). The reason is that the bit-field must be extracting from the rest of the bit-fields before it can be tested to see if it is true or false. Since a boolean value's primary purpose is to be testing for its value, this leads to poorer performance when using bit-fields.

This is a good place to point out that there are other types of optimization outside of performance and memory, such as network transfer speed. Although performance and memory are the most common types, these other types are important in certain cases and might require similar techniques and tradeoffs to those of performance optimization. For example, memory differences in the data that is sent over the network typically have a much greater impact than in other

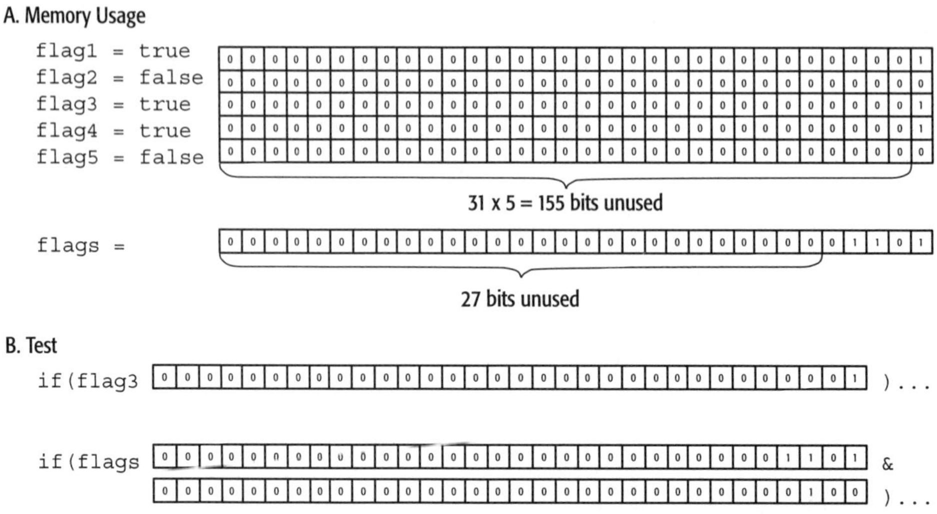

A. Memory Usage

flag1 = true
flag2 = false
flag3 = true
flag4 = true
flag5 = false

31 x 5 = 155 bits unused

flags =

27 bits unused

B. Test

if(flag3) ...

if(flags &
) ...

FIGURE 1.4   A) Example of the number of extra bits that go unused if a bit-field is not used. B) Example of the extra computation required when testing a bit-field.

cases. Storing boolean values as bit-fields would make sense if the full set of boolean values was to be transferred over the network on a regular basis. However, as with many exceptions, there is an exception to the exception. If the bit-fields are only sent occasionally, it might be better to interact with them locally as C++ booleans and convert them only when network transfer is needed. This complexity in the decision process about performance further emphasizes the need to wait until the code is solid at the end of development before optimizing, and to always test the optimizations for resulting performance changes.

Design

Changes in the application's design to improve performance can often be difficult to decide upon, but once the decision is reached it is usual easy to implement them since it often involves cutting or not implementing features. Unlike the rest of the optimization types, this optimization is rarely performed prematurely. Sometimes this optimization is not even performed when it is necessary, leading to missed deadlines, extra expense, and in some cases project cancellation. Therefore, you will rarely see this type as a premature optimization, and instead should look for the opposite case of resistance to change or remove features when approaching the end of development.

WHY PAST MISTAKES

One common occurrence that leads to our assumption that we know what is slow and how to take care of it early is our attempt to learn from past mistakes. Learning from failures is a very important part of the learning process, but this requires that we be able to distinguish the mistakes from events that are unpleasant but also unavoidable. Too often, after several projects where a large amount of optimization work had to be done at the end and required overtime and missed milestones, a programmer will decide that the appropriate way to handle this is to perform the optimizations as early as possible. This is the wrong way to handle this situation.

The problem was not that the optimizations were done at the end of the project, but in the method with which these optimizations were handled. Later in the "Prevention" section, we will talk about how to prepare for later optimizations, but not to make them until necessary. Otherwise, optimizing will actually take more time rather than less. The other important step than should have been made is to schedule time for these optimizations. This is the proper way to avoid falling behind on milestones and consequently the project finish date.

While this might all seem counterintuitive, first-hand experience has shown otherwise. A consultant once joined a project in its early stages. On all his previous projects, he had been brought in during the later stages of the project to assist in optimization and debugging. Convinced that one of the main reasons why optimization was so difficult on these projects, rather than poor coding practices, he decided to lay the groundwork for optimizing early and often in this new project. The result was a body of code that was difficult to modify but easy to break, and in addition an extra programmer was brought in at the end to help with optimization anyway.

Source Code

Source code optimizations do not improve application performance, but in theory save the programmer some typing. This type of optimization should never be done, but unfortunately, it is encountered all too often. There was once a day when this type of optimization was necessary because of the extremely limited memory and poor tools available on early computers. However, unless this book has been sent back through time, you should not be performing these optimizations anymore.

An extreme example of this type of optimization follows:

```
void hs(int *a, unsigned int n)
{

    typedef unsigned int ui;
    if(n<2)return;
    for(ui h=n/2-1,r=n;;){
    int v=r<n?a[r]:a[h];a[r<n?r:0]=a[0];
    ui i=h,j=2*h+1;while(j<r){
    if(j<r-1&&a[j]<a[j+1])++j;
    if(v<a[j]){a[i]=a[j];i=(j*=2)/2;}
    else break;}
    a[i]=v;if((h?--h,r:--r)==1)break;}
    a[0]^=a[1]^=a[0]^=a[1];
}
```

While you will never see anything this bad in real applications, it illustrates the difficulty this type of optimization can cause to the readability and ability to modify code. This function performs approximately the same operation as the heap sort function example given earlier in the discussion of readability, only now it is not difficult to read because it is an optimization to performance.

There is one use for this form of optimization: obfuscation contests. Most major languages have some form of this competition, and it can be quite an interesting learning experience to write one of these obfuscated contest entries. To achieve the necessary artistry in your entry, you must explore the details of the language and push it in directions that it would normally not be used. There are even lessons to learn that can be applied to real applications, just do not bring the obfuscated coding style along with them.

PREVENTION

The problem is determining when the optimization is required. The simple answer is at the very end of development, but this is not enough information for practical purposes on a real-world project. We must come up with better metrics for when to optimize if we are to prevent ourselves from optimizing too early. Without guidelines, the temptation will always be there, and in reality there are some cases where optimizations will be needed before the end of the project. Even more common is the need to avoid coding practices that make later optimizations difficult.

The Simplest Technique: Do Not Optimize

The simplest prevention technique is not to optimize. Resist the temptations, and wait until the end of development. Then, only optimize to achieve your performance goals; otherwise, you will be wasting time and money when the application could be in use already. This is an easy goal to say, but much harder to realize. The best tool at your disposal for achieving this goal is discipline. Think about every algorithm in terms of flexibility and clarity, and then decide whether you have accidentally considered performance. If performance was a motivating factor, revisit the other options and see if one of them is clearer and easier to modify. When possible, it is also very useful to have input from another programmer who is not directly involved with the algorithm in question. Just make sure that the other programmer understands the importance of avoiding premature optimization, and that you are looking for clear and robust code as opposed to performance-oriented code.

There is an exception to the rule of no premature optimization, but it should only be used after very careful consideration. If it is not possible to run and test the application due to extremely low performance, the only option might be optimization. In the "Cures" section, we will talk more about how to properly optimize the code even when it has to be done early. The important thing to remember about this exception is to take it very seriously, and to only optimize enough to continue development for a reasonable amount of time. Do not try to optimize up to performance specifications, or guess how much optimization is necessary to take you to the end of the project. Chances are that the guess will be wrong. Instead, estimate how much performance is necessary for another one to three months of development before optimization needs to be revisited. Often, this will last longer than expected, and you will not have wasted as much time on optimization that might be thrown away by the end of the project anyway. Again, always do this with great caution, or you could be creating problems for later development.

K.I.S.S. (Keep It Simple, Stupid)

Keep it simple, stupid. Not the most politically correct of statements, it nevertheless applies to many disciplines, including software engineering. Human beings have a penchant for making tasks more complicated than they need to be, and programmers are especially good at this. Therefore, while it might seem obvious that finding a simple solution is often the best course of action, it is necessary to keep reminding everyone of this to prevent straying from the proper course.

When it comes to optimization, keeping it simple means waiting until the end of development to add the complications inherent in most optimizations. Avoiding complicated code serves two purposes. The obvious effect is improved development speed and flexibility. Less obvious is the fact that simplicity can allow better optimizations than would otherwise be made. This follows from the fact that higher-level optimizations provide greater performance improvements on average than lower-level optimizations, and these higher-level optimizations are much easier to achieve if the underlying code is simple to understand and modify.

High-Level Languages

Computer languages create a large amount of controversy among software engineers, with many taking this almost to the level of a religion. While there are many valid arguments out there for this language or that language, one claim that never holds states that to truly optimize an application, it must be written in as low level a language as possible. This fallacy leads many to scorn the use of higher-level languages such as C++ and Java. Do not do this; you will be trading off development speed and flexibility for an uncertain gain in performance, and a more likely outcome of a canceled project.

The complexity of modern computer architectures provides another motivation for using higher-level languages. Many modern processors have reasonably large instruction sets and odd constraints caused by pipelining, timing, and other quirks. It would be impossible to train everyone on a team in the particulars of each processor they are to work with, especially if the team moves from one platform to

---

ASSEMBLY ONLY

One story involves a programmer evangelizing the use of assembly for the creation of applications. As proof, he offered a sound editing application written by him. The application performed well and had a reasonable number of features. Upon asking why more applications were not written in this manner, another programmer responded with a question of his own. He asked, "How long did it take to make this application?" The assembly programmer responded, "Eight years, why?" The simple fact was that an application that takes eight years to complete is extremely unlikely to be viable as a commercial application that will turn a profit. If your project has unlimited time and no concerns about money, you might be able to write it all in assembly, and even then, the complexity of writing assembly might prevent you from fully optimizing the application.

another on a regular basis. However, higher-level languages provide a solution to this by allowing an expert in the architecture to write a compiler that uses the higher-level information to optimize the low-level instructions generated during compilation. This frees the rest of the development community to focus on issues specific to their applications. Along with the benefit of hiding the complexity of a processor's architecture, the higher-level language allows the same code to be compiled on multiple architectures with only minor compatibility changes. With the common occurrence of multiplatform applications in recent years, this is a very important feature. Without this ability, it would be prohibitively expensive to develop for more than one platform.

Encapsulation and Abstraction

Surprisingly, one of the most important coding practices for proper optimization can have a negative impact on application speed. *Encapsulation* is the collection of related code into a single unit such as a class or module, which can then use *abstraction* to hide the details of the data and implementation behind an interface that is subjected to fewer changes than the internals (Figure 1.5). Object-oriented languages have language features that can enforce these design decisions. Even if the language you are using does not directly allow enforcement of this design, proper discipline and written standards can still be used to follow these practices. No matter which approach is taken, usually overhead is involved, primarily through the indirection that results from proper encapsulation and abstraction. If extra overhead results from using encapsulation and abstraction, how could these practices result in better optimization? The answer to this seemingly contradictory situation results from important optimization principles.

You do not want to optimize until you have determined what would most benefit from optimization. Proper encapsulation assists in this by allowing profiling information to show the correct location of slow code. Profilers, discussed in more detail later, have their own restrictions on performance and information detail. Organizing the code into discreet units allows the concentration of profiling efforts on the most likely targets for optimization.

Once an attempt at an optimization is made, the performance gain must be measured to determine if the desired result was achieved or if performance was actually reduced. Encapsulation allows for more focused optimizations and better tracking of the results of individual optimizations. Several different algorithms can be under consideration for an optimization, without a clear indication of which one would be most beneficial. Abstraction reduces the implementation time of testing multiple algorithms, thus providing an advantage in achieving the best optimization.

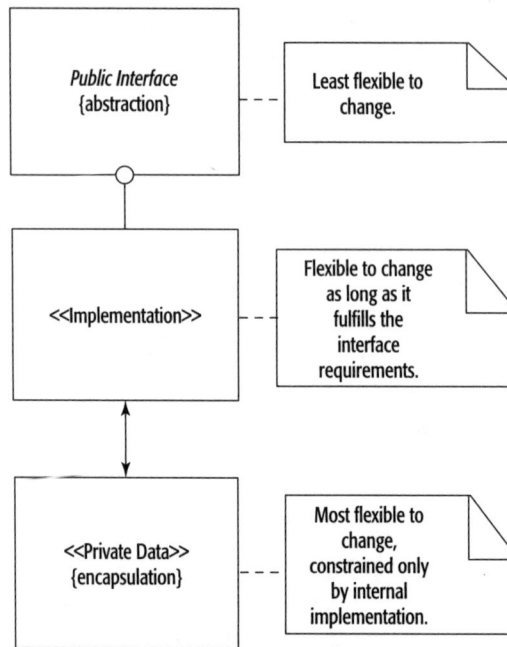

FIGURE 1.5   Relationship between encapsulation and abstraction.

Optimizations should be saved until the end of development as much as possible, but you still want to make as many optimizations as possible when the optimization phase does come along. Abstraction allows optimizations to be made only on encapsulated data and implementation without affecting the remainder of the code base. By limiting the amount of code that is required to change, more optimizations can be made without large costs in development and testing time.

Strategy Pattern

One of the most useful methods of encapsulation and abstraction for optimization purposes is the *Strategy Design Pattern* [GoF95]. The basic idea of the Strategy Pattern is to abstract the interface to an algorithm so that multiple algorithms for the same functionality can be encapsulated in interchangeable objects. This pattern is described in detail in the *Design Pattern* book by the Gang of Four, so here we will just examine how it applies to optimization.

Use the Strategy Pattern whenever possible for encapsulating any reasonable sized algorithm. This will aid greatly during optimization by allowing an algorithm to be switched by changing a single object creation call. By combining this with the *Factory Pattern* (Figure 1.6), which allows the creation of related objects without hard-coding the concrete class, several algorithms can be exchanged without re-compiling to test for the one that provides optimal performance. Furthermore, by limiting the interaction to the interface of the Strategy, the risk from changing algorithms is greatly reduced.

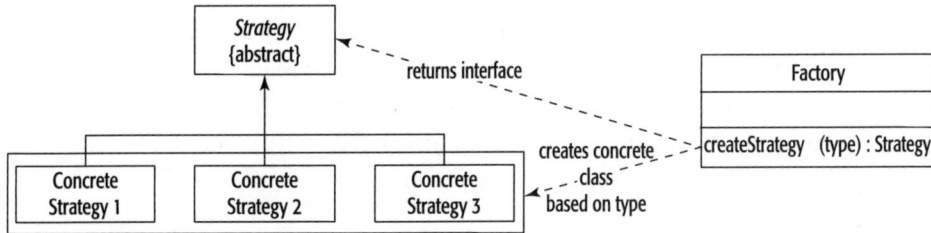

FIGURE 1.6 Use the Factory Pattern to create different concrete Strategy instances for testing. Each concrete Strategy implements the abstract Strategy interface and is therefore interchangeable.

Another potential benefit of the Strategy Pattern is the ability to dynamically change the algorithm that is used for a particular operation. This allows for the possibility of using different algorithms based on cues from the input data set in real time. While this technique is limited in the number of places it can be used, without using the Strategy Pattern it would not be possible at all.

Editors: Tools of the Trade

Earlier we discussed a form of optimization that should never be performed: optimizing the amount of text in the source code. While the reasons for avoiding this should be obvious, many programmers still fall into this trap because of the common human failing called laziness. Certainly, no one wants to do a lot of extra work that is unnecessary, but what appears to be laziness merely creates more work for us later.

Fortunately, a solution now exists that will allow us to maintain our laziness with only a small amount of up-front cost. This solution is to find one of the most important tools of the modern programmer, the editor, and learn how to use it to increase your code writing efficiency without sacrificing clarity. Here we will discuss several tools available in most of the powerful code editors available and how to take proper advantage of them.

One of the largest timesaving features that an editor designed specifically for your language of choice can offer is auto-completion of names. Auto-completion allows you to type only a portion of a name or statement and then press a single key or key combination to complete the rest of the name or statement. Ever year, the accuracy and speed of auto-completion technology is improving, leading to less and less typing while allowing longer and more descriptive names to be used. Pop-up lists are usually available to help you look for a name if you do not remember the exact name. Further pop-up comments can help to remind you what was the purpose of the code element. They are also becoming increasingly context sensitive, allowing the choices to be narrowed after fewer characters have been provided. One particular product, *Visual Assist* from Whole Tomato Software, even uses an intelligent guessing algorithm to provide a best guess before all ambiguities are resolved (Figure 1.7). Using this, you often only have to type one or two characters before clicking the Auto-Completion key.

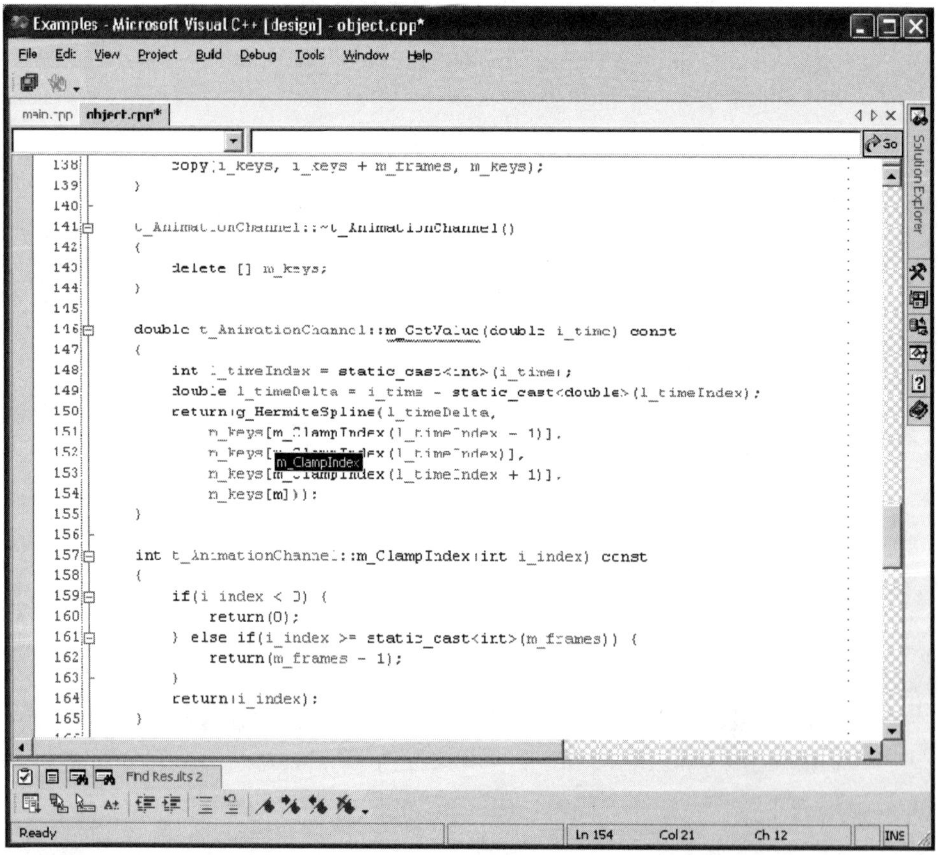

FIGURE 1.7   After typing only m, *Visual Assist* guesses from the surrounding context that m_ClampIndex is what the user wants. The user only needs to press ⨀ᴛᴬᴮ to complete the entire word.

Beyond simply using this feature, you can improve upon its efficiency by careful coding standards that help the editor decide among ambiguities faster. Try to differentiate the beginning of names as much as possible to allow you to type fewer characters to get to a unique name. Choosing a similar prefix for related names can also ease recall, especially if a pop-up list is available for your editor. Finally, if your language supports the separation of names into separate namespaces, separating names into the appropriate namespaces can reduce the clutter of names to choose from in a particular context. Figure 1.8 demonstrates how namespaces combine with auto-completion can save typing. If you are working in a team environment, an official naming convention is recommended to enforce this across the entire code base.

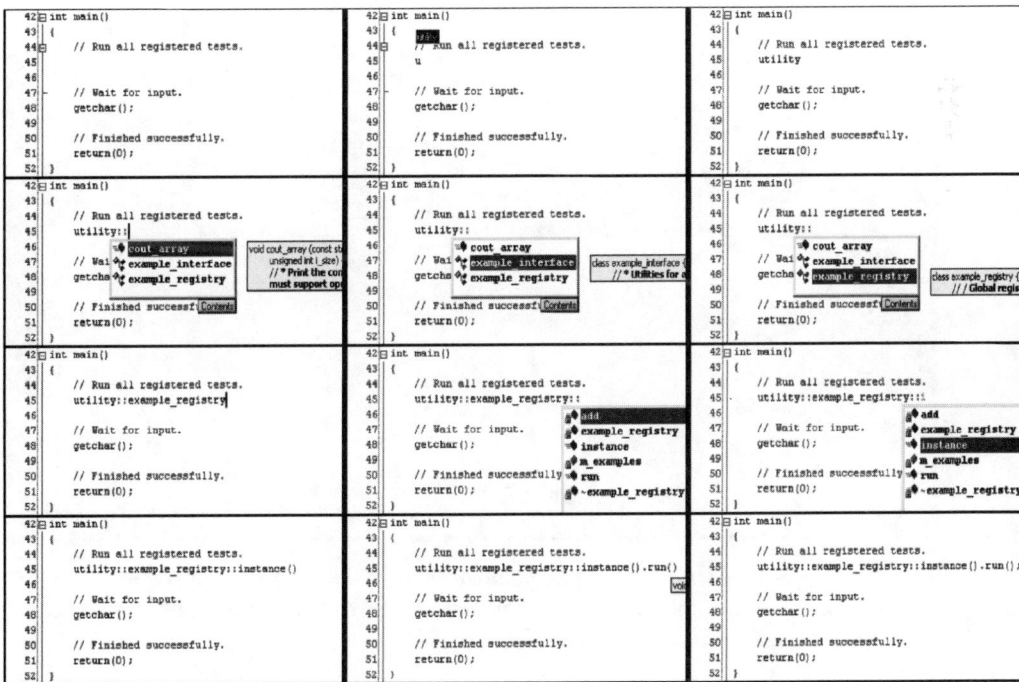

FIGURE 1.8 Using distinct namespaces and proper naming conventions, `utility::example_registry::instance().run();` can be entered into *Visual Assist* by typing only `u`[TAB]`::`↓↓[TAB]`::i`[TAB]`.run();`

Some editors also offer a feature that allows you to create your own shortcut character sequences that can then be used with the Auto-Completion key to provide user-specific code templates (Figure 1.9). This can save a lot of typing if you place your most commonly used statements in these templates and create shortcuts

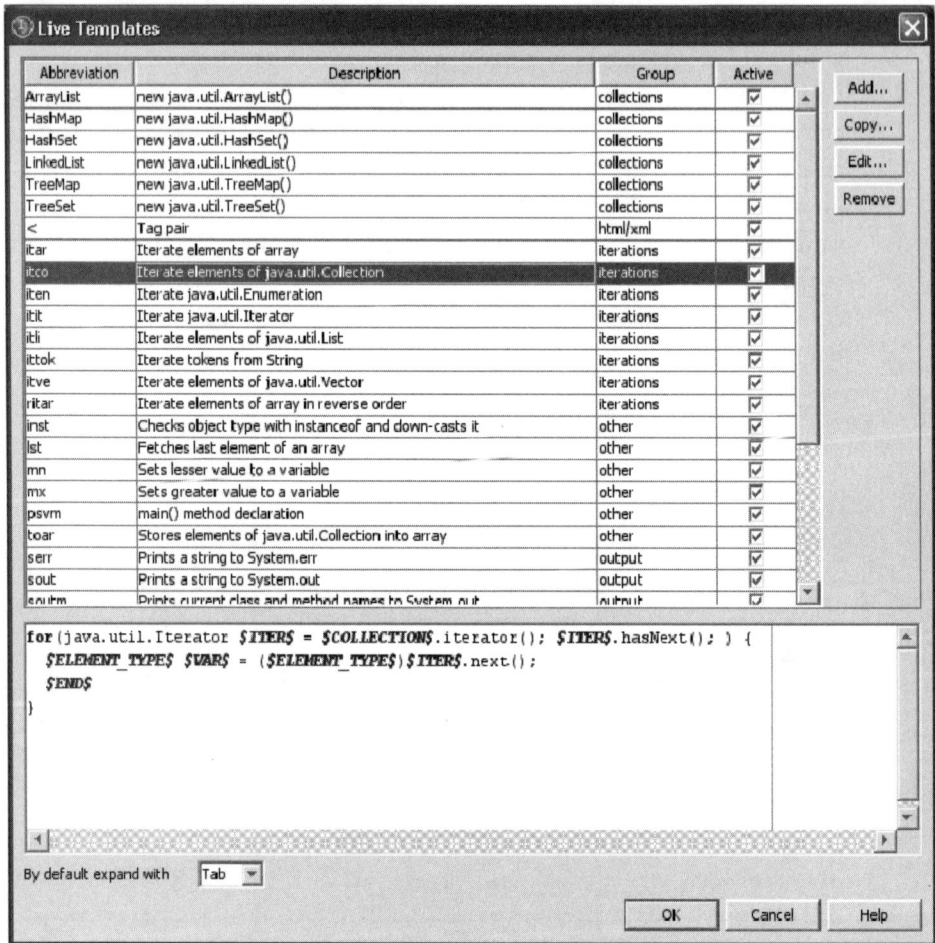

FIGURE 1.9 Special auto-completion sequences in IntelliJ's IDEA editor. Typing `itco` then the key sequence for auto-completion will expand to the text currently in edit box, where text surrounded with $ has special meaning.

that are meaningful to you. Be careful to choose shortcuts that will not interfere with other auto-completion names in your regular code base. File templates are also available in many editors to simplify creating new files and classes with less typing. These can be set to follow the coding standard of the team with correct copyright information and other standard layout conventions (Figure 1.10).

There are many other features available to specific editors that can help with code entry. Even if a feature is not available, all the major editors offer some form of extension language for adding your own functionality to the editor: Microsoft Visual C++ uses Visual Basic, SlickEdit uses SlickC, and Emacs uses a form of LISP.

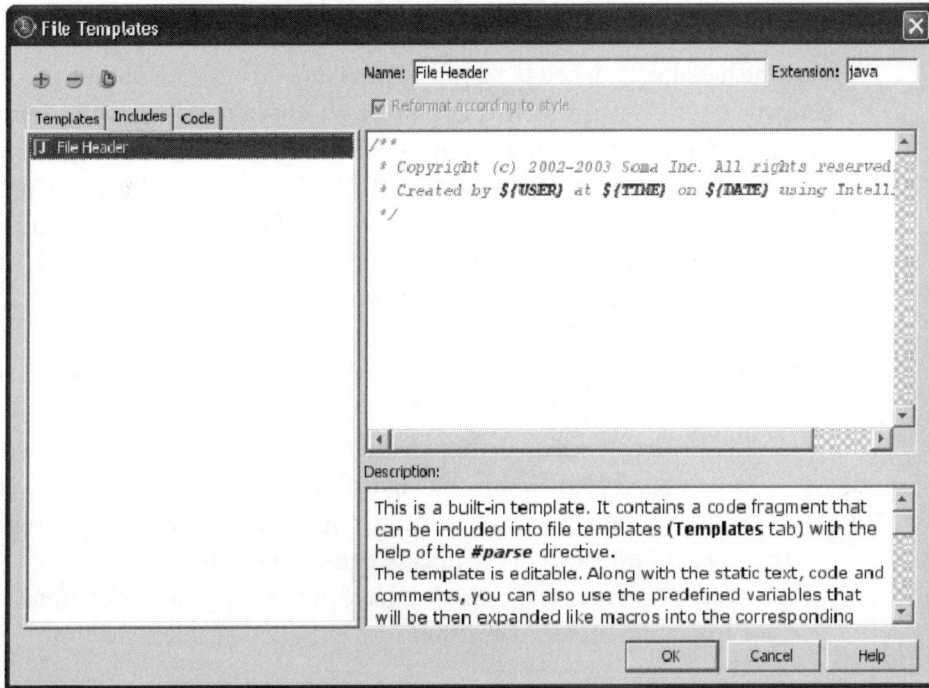

FIGURE 1.10 File template in IntelliJ's IDEA editor that places standard copyright information and file creation information in a comment at the top of new Java files.

Spending some extra effort writing utilities to perform common functions in your editor can save large amounts of production time throughout the current project and into future projects.

CURE

There are two parts to curing premature optimization. The first part explains what to do when inheriting code early in a project that has been optimized too soon. If this is not tackled immediately, time will be wasted over the course of development until the cure is implemented. Time wasted leads directly to money wasted, leading to a spiral of problems, so be sure to confront these issues as soon as possible. The second part is about knowing what to do when the optimization stage finally arrives. Without this knowledge, the optimizations will be misdirected and result in problems that are very similar to those of premature optimization.

When to Refactor

Someone has already taken the development time to write an optimized version of an algorithm, so why would there ever be a reason to rewrite it with a less optimal algorithm? For the answer to this, we look back to the symptoms of premature optimization. If the optimization is isolated and does not need to be changed, then there is very little reason to replace it. However, if the optimization is intertwined with code that requires a change, the difficulty in modification might be a reason for refactoring. In addition, this might bring future benefits if more changes become necessary.

Once the decision to refactor is made, the following steps should be taken:

1. Isolate the current algorithm from the rest of the code; reducing the other code's dependencies on the algorithm as much as possible. Refactoring should be done in small steps. It might require the introduction of additional abstractions in the interface to the algorithm.
2. Test to ensure that the changes have not broken the application.
3. Replace the current algorithm with a more readable algorithm.
4. Test.
5. Review the changes to make sure they have improved the readability and robustness of the code; preferably, have another programmer who has not worked on it review it as well.

If proper tests are not in place to verify that the code is not broken by the changes, now is a good time to add them. Do this before you refactor the code, thus providing regression test results from the current code to ensure that the new code provides the same functionality.

Profile, Profile, Profile

At some point, it will become necessary to optimize the application. For most projects, this should be near the end of development. However, on real-time applications and a few other types of applications, it might be necessary to perform some small optimizations during development to allow proper running of the application for testing and evaluation. Both types of optimizations are performed using the same techniques, but optimizations carried out earlier in the development cycle should only be performed until the application is returned to acceptable performance for further development. Final optimizations are required to be performed until the target performance required for application deployment is reached.

The most important tool required for performing these optimizations is the *profiler*. A profiler might be anything from small timing reports embedded in the code to a full separate application such as *Intel VTune Performance Analyzer* (Figure 1.11). The more profiling tools available, the easier it will be to get the data necessary for identifying the slowest code. An extremely useful feature to look for when considering commercial profiling tools is the level of support for *call graph profiling*. Call graph profiling allows the viewing of the entire call tree for profiling data and, as discussed later, is essential for determining the best optimizations. Another important consideration for real-time applications is the performance impact of the profiling on the application. This is particularly important when evaluating the performance of an application that requires constant user input. If the

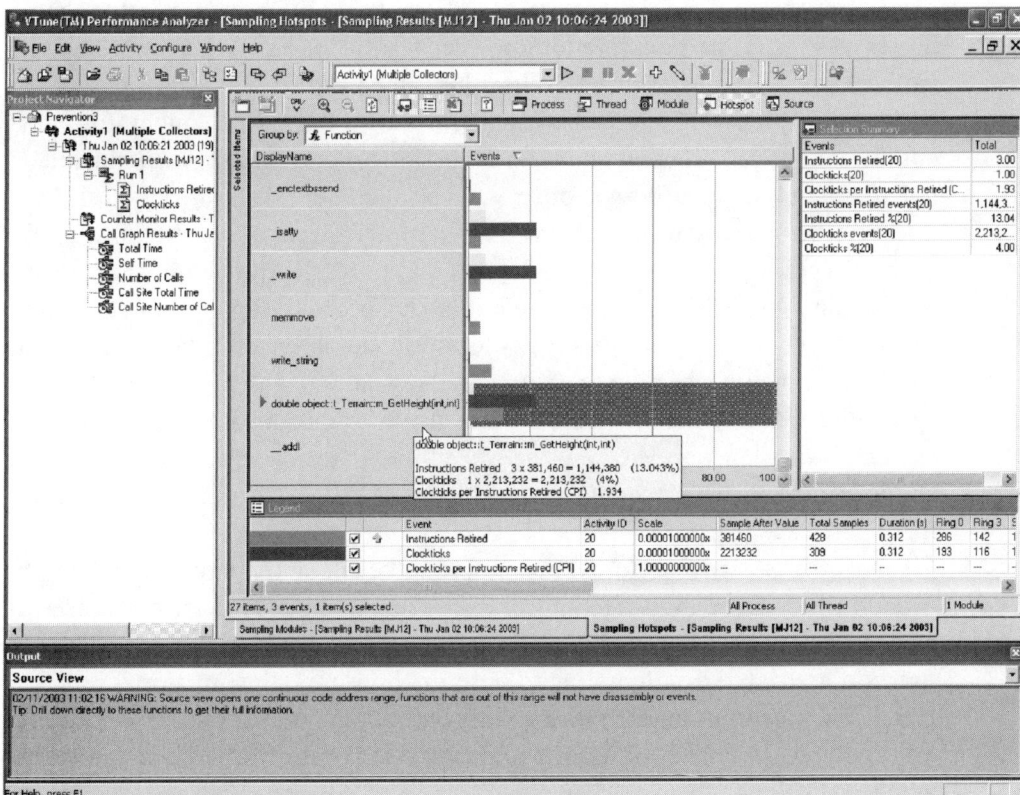

FIGURE 1.11 Intel VTune Performance Analyzer. Reprinted by permission from Intel Corporation, Copyright Intel Corporation 2001.

application becomes unusable due to poor performance, any profiling data collected must be suspect. This is another reason why a range of profiling tools is useful, allowing different types of profiling for different parts of the application.

Profiling and optimization is really an art, but this does not mean it should be treated as a chaotic process of random changes and uninformed guessing. There is a definite advantage to following a structured approach to optimization. What follows is a general plan for profiling and optimization that should be customized to meet the particular needs of your project.

First, use a profiling tool to determine which particular high-level module in your application is taking the most time. If the application is real-time, with or without user interaction, there are several considerations to take into account when performing this step. Real-time applications are generally profiled on a frame-by-frame basis. There are really two goals when trying to optimize this type of application: average frame time and frame time consistency. Tradeoffs might need to be considered to avoid spikes that are as noticeable if not more noticeable than a slower average frame rate. In addition, user interaction introduces multiple scenarios that need to be optimized. When this is the case, carefully separate and prioritize each scenario and tackle them one at a time. You might find that optimization in one scenario improves the performance of another scenario, but do not attempt to optimize both at once.

Once the performance concerns have been narrowed to a particular module, create a repeatable set of inputs to this module so results can be measured deterministically. While this might require minor changes to the application, the elimination of random results will easily pay for this in development time. For real-time applications, you might need to write special test cases to ensure this repeatability. Once this repeatable test harness is in place, use more detailed profiling tools to determine which function is taking the most time. If you have call graph profiling available, here is where it should be used. Be sure to record this information for future reference.

Now that a particular function is found to be taking the most time, the optimization that will be applied to improve performance must be determined. Here is where a mistake is often made. Instead of starting by attempting to optimize the function itself, start by attempting to optimize how often the function is called. A proper high-level optimization can greatly reduce the number of calls to the function, making low-level optimization of the function unnecessary. This is where call graph profiling is most valuable; allowing you to see which call tree is most responsible for the time taken by the function in question (Figure 1.12). Start look-

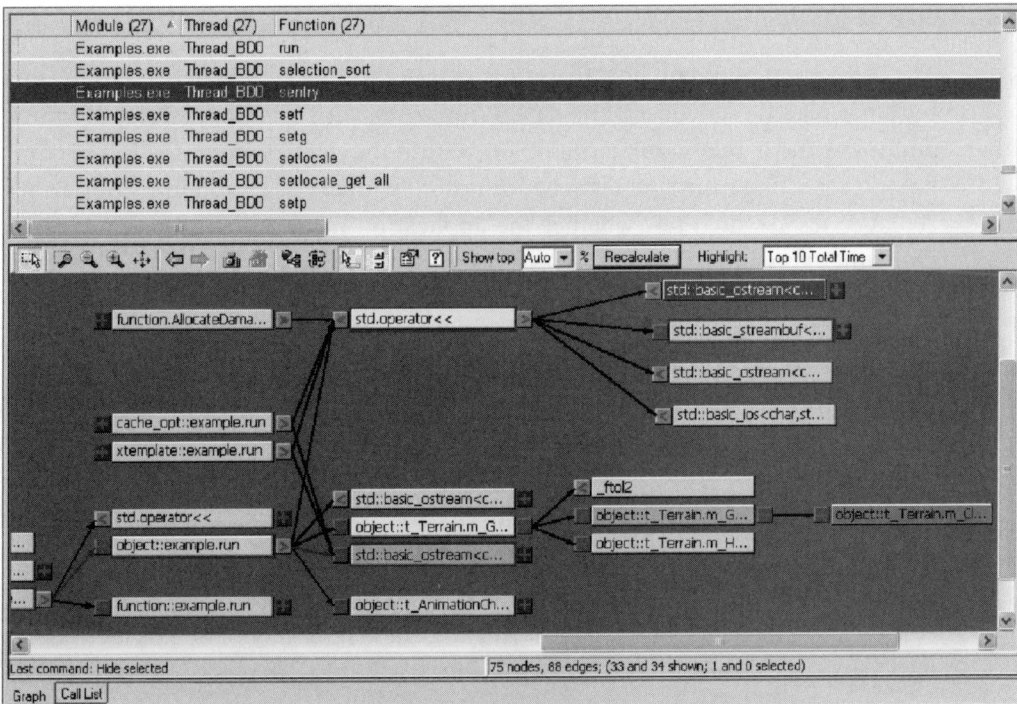

FIGURE 1.12 Browsing the call graph in Intel VTune Performance Analyzer can give more information on the source of expensive calls, providing extra context for deciding where to make optimizations. Reprinted by permission from Intel Corporation, Copyright Intel Corporation 2001.

ing for an optimization at the highest possible level, and work down to the lower levels only after determining that no optimization is possible at that level.

Once a single optimization has been chosen, implement the optimization. Then, run the exact same test scenario that was used to find the function to optimize. Compare the results, and ensure that the benefit of the optimization was worth the tradeoff in readability and robustness. If it was not substantial enough to warrant the changes, revert to the original code and search for another optimization.

Repeat this process until you have achieved the desired performance target. Be sure to profile after every code change during this process; otherwise, unexpected results might lead you to improper or useless optimizations. Do not be afraid to try several algorithms for the same optimization, but test each separately and compare the results of each with no other changes made to determine which one should be kept as the final optimization.

Importance of Testing

The plethora of books on how to perform proper testing is an indication of the importance placed on it. Several new methodologies, such as Extreme Programming, and many older methodologies rely on testing for their success. Testing encompasses everything from unit tests, which test individual classes and functions, to acceptance tests, which test whether the application meets the customer's criteria. One other very important set of tests occurs at the module and application level; these are integration tests used for checking the interaction between functions, classes, and modules. Some methodologies only use a portion of the possible tests; others can go overboard and require every single element to be tested. The best possible solution is a balance that aims to create tests for likely points of failure without wasting development time on tests that will never fail.

There are two very important aspects of testing as it relates to optimization. The first set of tests that are important are the normal tests used for development stability. These include the full range of test types normally used (Table 1.2), so they should already be available by the time the optimization phase is reached. After every change made to improve performance, these tests should be executed. If the tests pass, confidence that the optimization has not broken existing code will be

TABLE 1.2   Testing Types

Testing Type	Description	Scope	Automation
Unit	Testing to ensure that code obeys its contract and does what is expected	Single unit (class/function/etc.)	Full
Integration	Testing to ensure that code obeys its contract and does what is expected	Multiple units	Full
Error handling	Testing to ensure that errors are handled gracefully, including under heavy system load and full resource constraints	Single unit/multiple units/application	Full
User data	Testing with real customer data to ensure that the application works under real-world conditions	Application	Partial
Usability	Testing to ensure that the application is usable by the customer, largely user interface testing	Application	Minimal

high. This is extremely important at the end of the development cycle, when optimization should be performed, because even minor mistakes can cause large problems and critical delays.

Performance profiling tests are the other important set of tests. Often, potential optimizations can have no effect or a negative effect on performance. It is necessary to profile after every change to compare the performance with and without the change. If the change impairs performance, then reversion to the old code should be made. Even if there is no effect on performance, the reversion should be made, because any change presents a risk to the stability of the application. It would be unwise to take such a risk unless the gain is sufficient.

Creating profiling tests is easy for non-interactive applications, but a little bit more work is required to create repeatable tests for interactive applications. Figure 1.13 shows a basic setup for creating a profiling test that will work with an interactive application. While the application is running, the input data is collected. Once the conditions that the programmer wants to test are achieved, the data can be used to simulate the run repeatedly under similar conditions. Special consideration must often be taken in certain sections of the code to facilitate this type of testing. For instance, any randomizers must allow a repeatable seed to be provided so that random numbers remain identical across test runs. Simulation must also be able to function with timing information separate from the real-world timing. This allows changes to be made that affect the real-world timing while allowing the test runs to remain as identical as possible.

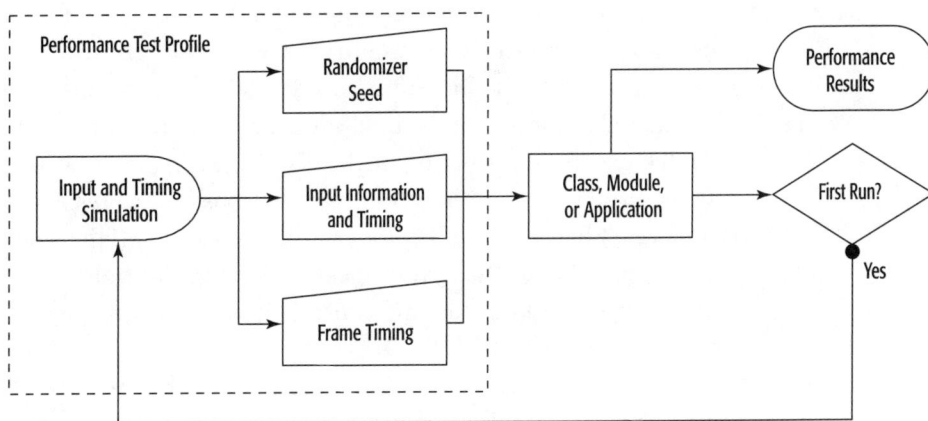

FIGURE 1.13 Profiling test setup for an interactive application. Inputs are stored and then played back to achieve runs that differ in as few ways as possible.

Notice that both cases stress the importance of performing the tests after every change. This comment deserves special note, as it can be tempting to make several changes at once and then perform the tests. Often this is because the testing process is time consuming. This can be reduced by ensuring that the tests are fully automated; essentially, one press should launch, and run to completion, the entire testing suite. The testing suites can also be broken into smaller test suites if it can be guaranteed that certain modules do not interact. With reasonably efficient tests, running the tests after every change will save a considerable amount of work when a particular change breaks the application. The cause will be evident immediately, and tracking down the cause is often the most time-consuming part of the debugging process.

Special Case: Libraries

Special consideration must be given to the optimization of libraries meant for external use. Unlike applications that go directly to the end user, libraries are a partway step to the final application. This leaves the library developer with less information about the end use of the library, making optimizations a trickier business. There are still methods that can be used to handle this case in a systematic fashion, leaving less to chance that the library users will come back complaining about how the library makes it impossible for them to optimize their applications.

The first step is to research the most likely uses of the library. If potential buyers are already known, consulting with them about their goals that would use the library can be of great use. The more detail you can get about what functionality they want and the framework within which they intend to use it, the easier it will be to properly optimize the library. If no potential buyers are ready at the time, which could be a bad sign for the salability of the product anyway, the intended frameworks within which the library will be used must be guessed at by the library development team. Once this information has been gathered, representative test cases must be created that represent several simplified versions of the uses of the library. This is a balancing act between the accuracy of the test cases and the length of development time used to create them. If proper development techniques were used, some of these test cases might already be available from the normal testing procedures used by the team.

The next step is to set up an automated profiling suite that includes all the test cases in the profiling. It is very important to consider all test cases for any one optimization; otherwise, only a small percentage of the library users will be happy with the performance. As in the optimization of a normal application, choose the highest performance impact to optimize first. This time, all test cases must be considered when choosing what to optimize. Once a change is made, profile all of the test cases again. Revert if any of the cases are negatively impacted by the change. It is important to make optimizations that benefit as many potential library users as possible, and it is even more important to avoid optimizations that help only one user to the detriment of other users.

RELATED ILLNESSES

Premature Optimization is the first major illness presented because it is the most prevalent reason for so many mistakes. Many of the other illnesses given later often have Premature Optimization as their starting point, including one of the other major illnesses: Not Invented Here (NIH) NIH Syndrome is caused by the belief that only you can possibly write code correctly, including code that performs to the level necessary. This belief stems from the incorrect assumption of Premature Optimization that everything must be optimal. Although there are other reasons for NIH Syndrome, understanding Premature Optimization eliminates a major reason, and puts you well on your way to preventing NIH Syndrome.

Another illness often driven by Premature Optimization is *Complexification*. As mentioned earlier, early optimization can be motivated by the desire to find a more interesting algorithm to work with rather than be bored with the simple algorithm. Avoiding Premature Optimization removes one more chance for Complexification to happen. Similar reasoning can apply to Hardcode, although that illness is often motivated by simple laziness.

Premature Optimization is often the cause of *Brittle Bones*, or fragile software architecture, as the dependencies and intertwined code can lead to an extremely fragile code base. The more optimizations that are made early in development, the more brittle the code becomes. This will soon build up and lead to major problems, usually at just the wrong time. Although avoiding Premature Optimization is only one step to preventing Brittle Bones, it is a very important step. Another causal relationship exists between source code optimization and i, or poor variable naming. Source code optimization is the primary cause of i.

On the reverse side, Myopia or shortsightedness is one of the causes of Premature Optimization. Short-term individual optimizations often get in the way of long-term design and full module optimizations. Remember to always take into account the long-term goals of the project, including performance goals, and consider how short-term decisions might cause problems for the longer term.

FIRST AID KIT

There are very few tools for preventing Premature Optimization. The primary tool in this case is your own skill and experience. Establishing the proper thought processes to handle developing code without Premature Optimization is the most important prevention technique. The other important tools are the proper high-level programming languages. With language features closer to the problem domain, there is less temptation to spend time optimizing low-level implementation details.

The effects of Premature Optimization, however, can be cured by several tools that assist the programmer. For example, when encountering prematurely optimized code while development is still in progress, you might need to refactor the code to improve readability and maintainability over optimization. Smalltalk's refactoring browser and the refactoring tools provided with IntelliJ's *IDEA* assist in this effort, automating the more tedious parts of refactoring.

In addition, when it does come time to optimize, there might be poor optimizations introduced by premature optimization. Whether or not poor optimizations have been introduced, the most important tool for the job of optimizing is the profiler. Profilers range from snippets of hand-written code to full software packages such as Intel's *VTune Performance Analyzer*. It is often necessary to avail yourself of several different profilers to reach the final performance goals of the application.

Both refactoring and optimization are assisted by two very important tools. The first tool type is testing tools. There are several freely available testing frameworks for different languages, such as JUnit for Java, CppUnit for C++, and PyUnit for Python. As with refactoring tools, these remove many of the tedious details from performing testing. Removing the boring aspects of testing is very important psychologically in encouraging testing, hence it is important to spend a little extra time learning to integrate these tools into your development and build processes.

The second type of tool is source control, such as *CVS, Perforce, or Microsoft's Visual SourceSafe*. The primary differences between the many versions of source control available are cost, features, and ease of use. Table 1.3 shows a comparison of three version control products as an example, but be aware that others are available. Determining which to use must be done on a project basis, based on budget, needs, and the skill level of those using the source control. Do not underestimate the importance of ease of use, because it can affect development time. Ensure that all developers who must use the source control chosen can use it efficiently.

TABLE 1.3 Comparison of Three Version Control Software Systems

	Perforce	*Visual SourceSafe*	*CVS*
License	Per user, volume discounts	Various	Open source
Interface	Command line/GUI	Command line/GUI	Command line*
Learning curve	Moderate	Short	Steep
Ease of use	Easy	Easy	Moderate
Flexibility	High	Low	Very high

* Separate GUI front-ends are available from other developers, most of them open source as well.

While many programmers have been told the importance of source control for the purposes of preventing the loss of code, less emphasis has been placed on its use during the process of refactoring and optimization. Particularly in the case of optimizations, and sometimes when refactoring, changes made do not have the desired effect. By using source control properly, the code state before the change was attempted can be stored and thus be available to revert to when a change does not work as desired. Figure 1.14 shows the workflow that takes advantage of this use of source control. This is an important part of the optimization cycle and can be extremely useful while refactoring as well.

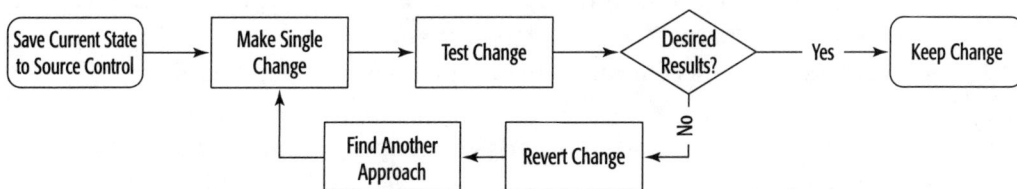

FIGURE 1.14 Workflow that takes advantage of source control for reverting failed changes.

To place these tools in context, here are the steps in a single optimization cycle and their associated tools:

1. Profile (profiler)
2. Modify (editor)
3. Profile (profiler)
4. Test (test framework)
5. Keep or revert (source control)

Each of these steps is important, and therefore the tools that are associated with them are a necessary part of the development process.

SUMMARY

Premature Optimization symptoms are:

- Most optimizations performed before the final stages of development.
- Poor readability and maintainability caused by optimized code.
- Unnecessary constraints imposed early in development.
- Poor optimizations performed early without the necessary knowledge of the complete application.
- The caching of data for optimization purposes too early in development.

To prevent Premature Optimization:

- Do not optimize until later in development.
- Keep code simple as long as possible.
- Use high-level languages.
- Use encapsulation and abstraction to prepare for later optimization, providing needed reassurance that optimization can be done later.
- Take advantage of editor tools to avoid source code that is hard to read.

To cure Premature Optimization and perform proper optimizations when they are called for:

- Do not be afraid to remove optimizations by refactoring to improve readability and maintainability.
- Profile often when optimizing.
- Test often when optimizing and refactoring.

2 CAP Epidemic

DESCRIPTION

Due to the prevalence of cut-and-paste (CAP) errors in the programming industry, this illness can be seen as an epidemic of huge proportions. The occurrences of development problems due to poor use of cut-and-paste is rivaled only by premature optimization in its impact, and in some organizations it exceeds everything as the primary cause for project failures. The driving force behind this is a misdirected sense of laziness, believing that it is easier and faster to use cut-and-paste than proper design and refactoring principles. While it might be easier and debatably faster at first, experience tells that it will cost time and money in the long term. Since software projects are generally several months to several years, the long term is very important, and cut-and-paste can cause a multitude of problems.

The good news is that cut-and-paste is easier to remedy than the more complex problem of premature optimization. Often, the largest obstacle is a lack of understanding of the detrimental impact that cut-and-paste has on software development and maintenance. Once the trouble is recognized, there are a multitude of tools and techniques available to help overcome the problem. Do not take this to mean that no subtle forms of cut-and-paste exist. A few instances of cut-and-paste are harder to recognize, which we will discuss later.

SYMPTOMS

Many of the symptoms of cut-and-paste are obvious, but it is still easy to overlook them. Of course, if you are the one doing the cut-and-paste, the symptoms are usually obvious when you select the cut operation and then apply the paste operation in your favorite text editor. Using these editor tools alone is not necessarily bad, which is why we will cover the various indications of the improper use of cut-and-paste along with a few subtle clues to instances of cut-and-paste that might otherwise be missed.

The Bug Came Back

Here is an all too common scenario that occurs while debugging an application. A bug is reported and assigned to a programmer. The programmer looks at the bug, and then replicates the tester's description to reproduce the bug. Once he reproduces the error, the code can be traced through and the location of the failure determined. The responsible code is then changed to fix the problem, and the bug is submitted as fixed. The tester can then reproduce the situation and mark the fix as verified. This should be the end of a successful resolution to a bug.

However, when cut-and-paste is used during development, several complications can arise after the problem is fixed. The most common occurrence is the reemergence of the bug with a different method of reproducing it. Even in the best-case scenario, this means that the programmer and tester will need to go through the entire fix-and-verify cycle again. This includes finding the exact location of the new error in the code, changing the code to eliminate this error, and then testing to ensure that the error was fixed without introducing new problems. While one small consolation is that knowledge from fixing the original error can reduce the time required to change the code, finding the error and testing that it is fixed are much more time consuming and do not benefit nearly as much from knowledge of the previous error.

Take the vector cross product as an example, where `vector1` and `vector2` are three-dimensional vector objects:

```
Vector cross_product = new Vector(
    vector1.y * vector2.z - vector1.z * vector2.y,
    vector1.x * vector2.z - vector1.z * vector2.x,
    vector1.x * vector2.y - vector1.y * vector2.x);
```

This is only necessary for a certain vector calculation, so a programmer might decide to use it directly in the code without adding it to the vector class. However, another programmer comes along later and finds another location that needs the functionality. Not sure if he wants to modify the vector class, he cut-and-pastes the code into the new location. This might happen a few more times before it is time to test the application.

Testing begins, and the following bug is reported:

```
Beam graphics appear to be distorted in
appearance.
```

After a couple hours of searching, this leads the programmer to find a bug in his code:

```
// ...
Vector beam_up = new Vector(
    beam.y * camera.z - beam.z * camera.y,
    beam.x * camera.z - beam.z * camera.x,
    beam.x * camera.y - beam.y * camera.x);
// ...
```

This is promptly changed to:

```
// ...
Vector beam_up = new Vector(
    beam.y * camera.z - beam.z * camera.y,
    beam.z * camera.x - beam.x * camera.z,
    beam.x * camera.y - beam.y * camera.x);
// ...
```

The code is submitted, retested, and verified as correct. However, this code still exists in several other places across the application. Another bug report is guaranteed to occur, and the debugging process will begin again. Perhaps another programmer fields this report, and spends an hour to find the bug in his code:

```
// ...
Vector spark = new Vector(
    object1.y * object2.z - object1.z * object2.y,
    object1.x * object2.z - object1.z * object2.x,
    object1.x * object2.y - object1.y * object2.x);
// ...
```

Then fixes it:

```
// ...
 Vector spark = new Vector(
    object1.y * object2.z - object1.z * object2.y,
    object1.z * object2.x - object1.x * object2.z,
    object1.x * object2.y - object1.y * object2.x);
// ...
```

This process will be required for each instance in which the original code was cut-and-pasted. Each occurrence will waste development time that could have been used on work that is more productive, or allowing an earlier completion time to be achieved.

That is not the end, however, as things can get even worse. The programmer might ignore the bug thinking that he has already fixed it, not realizing that another occurrence of the code exists. In the earlier example, this could occur easily if there were two different types of beams and another programmer had created the second beam type by cut-and-pasting the code from the first beam type. Alternatively, the tester might see it as a reoccurrence of the original bug and resubmit the original bug report without updating the procedure to reproduce the problem. This could also easily occur in the case where there are two separate occurrences of code for displaying beams. Worse, the new bug could be assigned to a different programmer who does not even have the knowledge of the original fix to help him. Moreover, when cut-and-paste is prevalent in an application, the bug could reoccur several more times with different methods of reproduction. Repeatedly fixing the same bug is an obvious sign of cut-and-paste that makes the already difficult debugging process exasperating.

Search and Replace

A direct consequence of cut-and-paste is the use of search-and-replace for making changes to the code that has been created with cut-and-paste. While there are other uses of search-and-replace, if you find yourself using it often on a project, chances are good that cut-and-paste is also in heavy use. One of the major problems of the basic search-and-replace functionality is the automation with which it operates. Normally automation is extremely useful for improving development speed and efficiency, but in this case, the programmer could overlook changes that are invalid or problematic because the process is automated.

When search-and-replace is used, make sure that each instance is carefully examined to ensure that the results are those desired. Do not use the replace all feature even if it is available, as this will blindly replace code without your knowledge; This is like closing your eyes and using a machine gun to try to save a hostage; chances are high that more harm will be caused than good. However, it is equally important to determine why search-and-replace is necessary. If the root cause is cut-and-paste, then code should be refactored rather than just replaced. Even if cut-and-paste is not involved, there might be refactoring tools, which we will discuss later, to help.

What Does That Mean: CAP Symptoms in Documentation

Poor documentation is an illness in its own right, but a particular type of documentation error can indicate the presence of cut-and-paste code. Have you ever read a comment that seems to go with the code below it, except that it gets one fact blatantly wrong, or sometimes it doesn't even seem to relate at all to the code below it? Chances are that you might find a similar but modified version of the code and an exact duplicate of the comment elsewhere in the code base. Programmers often cut-and-paste code and modify it slightly without updating the comments. Of course, the code does not have to be cut-and-pasted for the comment to get out of date. Searching for a distinguishable part of the comment elsewhere in the code can quickly resolve this question, and it is worth it to find cut-and-paste code that requires refactoring.

To illustrate this, let us start with the following code:

```
// Compute the average of the current frame
// and the next frame.
float midframe = (frame[current] +
    frame[current + 1]) / 2;
```

Now, this code could be cut-and-pasted to a new section and then slightly modified:

```
// Compute the average of the current frame
// and the next frame.
float midframe = (frame[current] +
    frame[current - 1]) / 2;
```

Notice that the comment was not updated along with the change to the code. This can easily lead to numerous problems, especially if the difference between the code and comment is subtle. Another programmer reading this could easily take the comment at face value and use this code with misleading expectations. However, by searching for the comment, the original code can be found and a comparison will easily show that the code differs. From this, it directly follows that the code must be looked at more closely.

Obscure Bug Hunt

Most of us have tracked down off-by-one errors, memory overwrites, and other mistakes caused by one wrong element in thousands or millions of lines of code. These tiny errors are insidious, causing frustration and hours of lost time. Cut-and-paste makes it easy to make such an error, leading to obscure and hard to track bugs often long after the programmer who wrote it has forgotten how that section of code works. So, how does cut-and-paste contribute to this problem? Let us go over a typical cut-and-paste scenario to see what can happen.

A programmer discovers the need for some new functionality that is very similar to another section of code that he knows about. Looking at the code, he determines how it can be modified to fit his needs and proceeds to cut-and-paste it into his own code. Just then, he notices another change that should be made. Since the change is quick, he goes ahead and makes it. Now he compiles the code and runs one simple test case that works as expected. Finished with that task, he moves on, having placed a subtle time bomb in the application. He forgot to make the change to the new code that he cut-and-pasted and, as luck would have it, the test case did not test for that change.

To show how easily this can happen, just consider code to determine the midpoint of a line. Start with the computation for the x coordinate:

```
midpoint.x = (current_line.endpoint1.x +
    current_line.endpoint2.x) / 2;
```

Now, cut and paste:

```
midpoint.x = (current_line.endpoint1.x +
    current_line.endpoint2.x) / 2;
midpoint.x = (current_line.endpoint1.x +
    current_line.endpoint2.x) / 2;
```

And begin to modify to compute the y coordinate:

```
midpoint.x = (current_line.endpoint1.x +
    current_line.endpoint2.x) / 2;
midpoint.y = (current_line.endpoint1.x +
    current_line.endpoint2.y) / 2;
```

Oops, the telephone rings. After answering the telephone, you move on to the next section of code. Unfortunately, you have left a small error waiting to appear once testing begins. This error is particularly insidious if proper unit testing is not being done, leaving the error to be discovered much later.

The more code that is copied during a cut-and-paste operation, the more likely a small change will be forgotten. The more cut-and-paste that is performed, the more likely these tiny errors will creep into the code. Writing proper tests before copying the code can help somewhat, but that is more like treating the symptoms rather than the disease. It is easy to miss something when creating test cases, and cut-and-paste requires more test cases than properly written code does.

Too Many Directions

Imagine these instructions to create a new build of an application:

1. Make changes to the code and/or assets.
2. Update the version number in the manifest file.
3. Update the version number in the application description file.
4. If image tile assets have changed, update the width and height of the images in the source code.
5. If map assets have changed, update the map size in the source code.
6. Build and verify.

A short list, but if you look closely, there are three extra steps that should not be necessary. Each of these steps is a chance for human error to occur and wreck the build. The reason for these three extra steps lies in the build design, which forces cut-and-paste updating of the application. For example, the build process should automatically update the version number in the application descriptor file based on what is in the manifest file. Likewise, the assets already contain information on their size that should either be read programmatically at run time or preprocessed by the build process. These changes would result in a much shorter list of actions:

1. Make changes to the code and/or assets.
2. Update the version number in the manifest file.
3. Build and verify.

This is a symptom of forced cut-and-paste programming that often appears later in development or when reusing code from a previous project. Eliminating the instructions that require cut-and-paste should be top priority on any project. Risk is greatly reduced by removing human interaction from repetitive tasks. Now some of you might believe that you can assign a less expensive employee to handle these repetitive tasks and avoid losing money that way, but this fallacy falls flat in practice. This person will likely become a bottleneck for making changes and is still likely to make mistakes that break the application. Always remember, a little time up front will save a lot of time later.

Information Duplication

It is important to remember that cut-and-paste problems apply to more than just code. In addition, cut-and-paste is not the only method of creating duplicate information. It is easy for two programmers to create similar pieces of code on the same project without realizing it just because of poor communication. What we are really talking about is removing duplicate information from human interaction. Any time there are two occurrences of similar code, there is a potential for error. Later, we will talk about techniques for fixing these instances in already existing code, but first we will examine how these poor coding practices can be avoided all together.

PREVENTION

Preventing the simplest form of cut-and-paste is relatively easy, avoid using the cut-and-paste feature of your editor whenever possible. However, to avoid all of the problems that the cut-and-paste epidemic implies, we must be cognizant of other forms of duplication as well. The complete goal of preventing cut-and-paste programming can be summed up in two rules:

- There should be one and only one copy of any human editable information.
- Any human editable information should not be derivable from other sources.

These rules encompass not only using cut-and-paste, but also preventing the creation of circumstances that force the use of cut-and-paste to make changes. This is an important risk management investment, as it will greatly reduce the chances for needless human error. Now let us examine ways that we can accomplish this task beyond the simple avoidance of cut-and-paste.

Know Your Code

The first step to reducing the amount of duplicated code within a project is to know what other code is written or being written for the project. Code reuse is essential to avoiding duplicate code. An important piece of achieving this goal is proper communication with other programmers. Understanding the general area that everyone is working on will allow you to direct queries about existing functionality to the correct programmer quickly and easily. On small teams, communication is probably the most efficient method for discovering existing functionality within the project's code base.

Extreme Programming attempts to maximize the level of communication using a technique called *pair programming*, where two programmers work on the same code together, with one writing the code and the other offering advice. With pairs of programmers switching partners on a regular basis, understanding of the code base disseminates quickly through the team. Extreme Programming also suggests regular code refactoring, which we will see later is a cure for the problems of cut-and-paste. Even if you do not want to engage in pair programming, keeping close to the other programmers you are working with will greatly increase everyone's productivity.

On larger projects, particularly projects where programmers are not centrally located, there is a need for tools to facilitate communication about what code already exists. Even though these tools are not essential to smaller projects, they can still be beneficial.

The first and most important tool is *automated documentation*, which extracts the documentation directly from the current code and comments. Later when we talk about docuphobia we will go into detail about the best practices for proper documentation, but here we will take a closer look at how automated documentation in particular can benefit code understanding and reuse. When we talk about automated documentation tools, we are discussing tools such as Doxygen and Doc-O-Matic that extract code documentation from the structure of the code and the comments associated with the code. These produce documentation that can be easily browsed or searched to find code that you might need. More information on these tools is available in the *First Aid Kit* for Docuphobia. Once found, you can reuse the code directly and avoid duplicating both the work and the code yourself.

Compiling of automated documentation is best done with a build of the code on a daily basis. This ensures that the documentation matches a currently working build. Otherwise, if any errors occur within the build, they can be resolved before the documentation is updated. There are also tools for checking to ensure that the documentation matches the basics of the code to which it is associated. Although this cannot correct for errors in description, it can point out missing or syntactically incorrect documentation. More information on one such tool, iDoc, can be found in the *First Aid Kit* for Docuphobia. These documentation errors should be included with the build report and marked for timely resolution. Preferably, this documentation is then available on a local HTTP server for easy access by all team members.

Another useful set of tools to improve your understanding of the code base you are working with is *code visualization* tools. These provide tools for visual representation of class hierarchies, relations, and other important aspects that are normally difficult to see from the code alone (Figure 2.1). They also commonly provide advanced search tools for quickly locating code with various syntactic properties. Although this cannot take the place of well-documented code, it can be a useful addition to existing documentation, or a fallback when you are stuck with poor to no documentation.

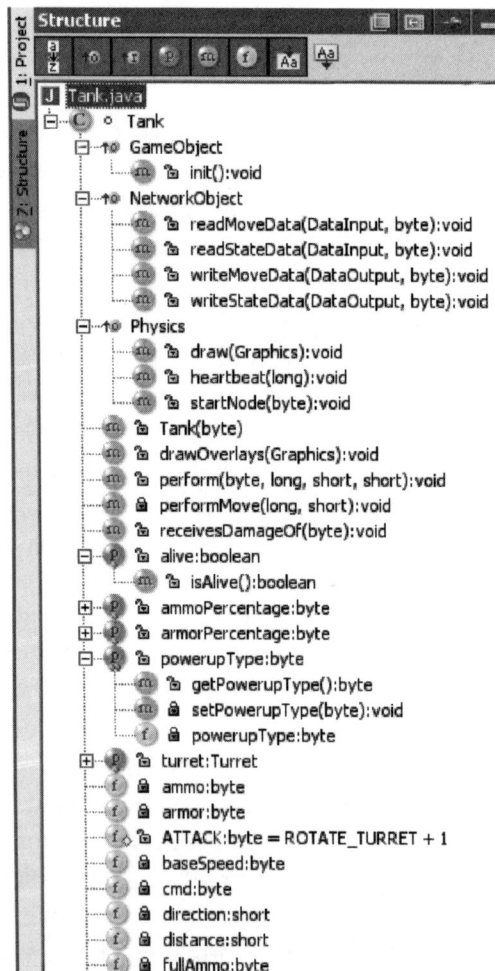

FIGURE 2.1 The Structure window in IntelliJ's IDEA allows the user to visualize the structure of a class, including where an overridden method originates and what accessor methods are present.

Know Your Libraries

One sure way to avoid writing duplicate code is not to write the code at all. However, what does that really mean? If you need the functionality, how can you avoid writing the code that accomplishes that functionality? The solution is extremely simple, but often missed: use third-party code or libraries. If the code to solve your problem has already been written, then reusing that code will avoid the possibility that you will be writing duplicate code since you didn't write any code at all.

In addition, libraries might provide new and clever methods for avoiding code duplication that you have never seen before. To illustrate this, we will look at a C++ library called *boost* that provides advanced functionality and techniques supported by experts in the C++ language community. Although not a part of the standard library, many proposals for the next revision of the standard library are derived from work done on boost. First, let us look at a common goal when creating a certain type of object in C++. The basic idea is to prevent the object from being copied or assigned, because this operation does not make semantic sense for this object. To accomplish this, we declare, but do not define, the copy constructor and assignment operator as private members of the class. This prevents access to this functionality except from within the class, and by not defining the copy and assignment functions a link error will occur if they are accessed within the class. An example of such a class is as follows:

```
class t_DoNotCopy
{
    // ...

private:

    t_DoNotCopy(const t_DoNotCopy &);

    const t_DoNotCopy &operator=(
        const t_DoNotCopy &);

};
```

This solution is elegant, but it must be duplicated in every class for which copy is to be disabled. The boost library provides a more concise solution to this by defining the following class:

```
// boost utility.hpp header file
```

```
//   (C) Copyright boost.org 1999. Permission to copy,
//   use, modify, sell and distribute this software
//   is granted provided this copyright notice
//   appears in all copies. This software is
//   provided "as is" without express or implied
//   warranty, and with no claim as to its
//   suitability for any purpose.

//   See http://www.boost.org for most recent
//   version including documentation.

//   class noncopyable

//   Private copy constructor and copy
//   assignment ensure classes derived from
//   class noncopyable cannot be copied.

//   Contributed by Dave Abrahams

class noncopyable
{
protected:
    noncopyable(){}
    ~noncopyable(){}
private:
    // emphasize the following
    // members are private
    noncopyable( const noncopyable& );
    const noncopyable& operator=(
        const noncopyable& );
}; // noncopyable
```

By deriving privately from this class, the amount of code that must be duplicated for each class is greatly reduced. The new version of the class given earlier now shows the savings that this library introduced:

```
class t_DoNotCopy : boost::noncopyable
{
    // ...
};
```

There is always a caveat. Actually, there are several that we will look into in detail when we talk about NIH (Not-Invented-Here) Syndrome, which deal directly with the reasons for code reuse. However, one of these caveats deserves particular mention here as it relates directly to the duplication of code. Do not use library functions or third-party code from two different sources that accomplish the same task. This will make your code less readable, and require more work in maintaining it when any of the libraries change. Be sure to evaluate and decide which libraries to use before writing the code so that the libraries can be used consistently across the application.

High-Level Languages

At the dawn of computers, programs were entered one byte at a time directly into a form the machine could process. There was no choice but to write every bit of code that was needed, even if that meant duplication. Time has changed all this as computer languages continue to evolve in a direction that reduces the effort needed to create applications. Many of these advances allow code to be written in a more generic fashion so that code duplication, and hence cut-and-paste, can be reduced or eliminated altogether. The following sections examine the most common language features that reduce code duplication. Most of them are present in several different modern computer languages, but the examples will be drawn from only one language at a time for clarity.

Functions

One of the earliest and most fundamental concepts to be introduced to higher-level programming languages was the *function*. Functions provide a simple means for reducing code duplication by allowing the parameterization of functionality for easy reuse. Functions exist in some form in almost every programming language, and it is a certainty that every programmer understands the basics of their use. Despite this, functions are often not used to full advantage in many applications. Therefore, let us take a deeper look at how functions prevent cut-and-paste and why they should be used more often.

One common problem with the way many programmers write functions is in the granularity of the function. Functions often become stuffed with several responsibilities, hence growing large and unreadable. There are several reasons to avoid this and write small functions with clear responsibilities. The functions become immediately more readable, and their purpose is easier to grasp. Documenting the function is easier and therefore likely to be more accurate. With better documentation, it becomes easier for both you and other programmers to find and reuse the function. This chain of reasoning should make it clear that breaking code into functions with well-defined responsibilities encourages code reuse and avoids duplication. So, what are the arguments against it? The most common argument is performance, as most language implementations of functions have some overhead associated with each function call. However, in Chapter 1, "Premature Optimization," the reasons that this should not be a concern during development are discussed in detail. It is preferable to write code that is easier to read, maintain, and reuse until optimizations are necessary.

The other common reason for not breaking code down to the proper number of functions is laziness. Programmers often do not want to spend the time creating new function definitions. This is a misplaced sense of laziness, however, since in reality this creates more work later. In addition, modern editors contain many tools to help ease the amount of work involved in writing code. What this really means is trading a little more time initially to save time later, both in maintenance time and by the fact that it will be easier to avoid writing duplicate code.

To make this more concrete, let us look at a sample function that is written poorly and then see how it should have been written. For this example, which can also be found on the companion CD-ROM in Source/Examples/Chapter2/function.cpp, we will need the following definition of an object that can be damaged and some properties for that object:

ON THE CD

```
/// Information on an object that can be damaged.
/// Most objects with require two copies of this
/// object to exist, one storing the base
/// values and one storing the current values.
struct DamagableObject
{
    /// Current hitpoints.
    unsigned int hitpoints;

    /// Current attack power.
    unsigned int attackPower;

    /// Current defensive power.
    unsigned int defensivePower;
};
```

For simplicity it is represented as a structure upon which our function will act, but later we will describe how proper function use is important in object-oriented programming as well. Next, we look at the function that performs damage allocation to an object:

```
/**     Apply effects from damage to an object.
 *      @param    io_object - current object state
 *      @param    i_base - original object state
 *      @param    i_damage - number of damage points
 *                    to apply
 *      @return   true if damage destroys object,
 *                false otherwise
 *      @notes    Single monolithic version of function.
 */
bool AllocateDamage_BeforeRefactoring(
    DamagableObject &io_object,
    const DamagableObject &i_base,
    unsigned int i_damage)
{
    // Check for destruction.
    if(i_damage > io_object.hitpoints) {
        io_object.hitpoints = 0;
        io_object.attackPower = 0;
        io_object.defensivePower = 0;
        return(true);
    }

    // Adjust hitpoints.
    io_object.hitpoints -= i_damage;

    // Update stats.
    io_object.attackPower =
       (i_base.attackPower *
       io_object.hitpoints) / i_base.hitpoints;
    io_object.defensivePower =
       (i_base.defensivePower *
    io_object.hitpoints) / i_base.hitpoints;

    // Apply damage visuals.
    if(io_object.hitpoints < 1) {
        cout << "Damage Effect: Explosion" << endl;
    } else
    if(io_object.hitpoints <
```

```
            (i_base.hitpoints / 4)) {
            cout <<
            "Damage Effect: Heavy Smoke and Fire" <<
            endl;
        } else
        if(io_object.hitpoints <
            (i_base.hitpoints / 2)) {
            cout <<
            "Damage Effect: Heavy Smoke" << endl;
        } else
        if(io_object.hitpoints <
            ((i_base.hitpoints * 3) / 4)) {
            cout <<
            "Damage Effect: Moderate Smoke" << endl;
        } else
        if(io_object.hitpoints < i_base.hitpoints) {
            cout <<
            "Damage Effect: Light Smoke" <<
            endl;
        }

        // Object not destroyed.
        return(false);
    }
```

Notice the use of comments within the function. These are necessary to understanding the workings of the function, and even so, it requires a closer read to determine everything the function is doing. This is because the function has assumed several responsibilities by itself: adjusting hit points, adjusting attack and defensive power, applying visual effects, and determining if the object is destroyed. We can instead write each of these responsibilities as a separate function:

```
/**   Modify the hitpoints of an object based on
 *    damage points.
 *    @param   io_object - current object state
 *    @param   i_damage - number of damage points
 *                   to apply
 */
void AdjustHitpoints(DamagableObject &io_object,
    unsigned int i_damage)
{
    if(i_damage > io_object.hitpoints) {
        io_object.hitpoints = 0;
```

```
    } else {
        io_object.hitpoints -= i_damage;
    }
}


/**    Update the attack and defensive powers of
 *     an object based on its current state.
 *     @param    io_object - current object state
 *     @param    i_base - original object state
 */
void UpdatePower(DamagableObject &io_object,
    const DamagableObject &i_base)
{
    io_object.attackPower =
        (i_base.attackPower *
        io_object.hitpoints) / i_base.hitpoints;
    io_object.defensivePower =
        (i_base.defensivePower *
        io_object.hitpoints) / i_base.hitpoints;
}


/**    Update the visual appearance of the object
 *     based on current hitpoint level relative to
 *     original hitpoint level.
 *     @param    i_object - current object state
 *     @param    i_base - original object state
 *     @notes    Prints damage effect to standard out
 *               as a test.
 */
void UpdateVisualDamage(
    const DamagableObject &i_object,
    const DamagableObject &i_base)
{
    if(i_object.hitpoints < 1) {
        cout << "Damage Effect: Explosion" << endl;
    } else
    if(i_object.hitpoints < (i_base.hitpoints / 4)) {
        cout <<
        "Damage Effect: Heavy Smoke and Fire" <<
        endl;
    } else
    if(i_object.hitpoints < (i_base.hitpoints / 2)) {
        cout <<
        "Damage Effect: Heavy Smoke" << endl;
```

```
    } else
    if(i_object.hitpoints <
        ((i_base.hitpoints * 3) / 4)) {
        cout <<
        "Damage Effect: Moderate Smoke" << endl;
    } else
    if(i_object.hitpoints < i_base.hitpoints) {
        cout <<
        "Damage Effect: Light Smoke" << endl;
    }
}


/** Check to see if an object is destroyed.
 *    @param     i_object - current object state
 *    @return    true if object is destroyed,
 *               false otherwise
 */
bool IsDestroyed(const DamagableObject &i_object)
{
    return(i_object.hitpoints < 1);
}
```

Notice the improved clarity provided by the function comments that were not possible before. The damage allocation function did not and should not include this level of detail in its comments because its functionality could change. Now we can rewrite the original function in a much clearer form:

```
/** Apply effects from damage to an object.
 *    @param     io_object - current object state
 *    @param     i_base - original object state
 *    @param     i_damage - number of damage points
 *                   to apply
 *    @return    true if object is destroyed,
 *               false otherwise
 *    @notes     Refactored function uses several
 *               smaller functions to accomplish
 *               its task with greater readability
 *               and reusability.
 */
bool AllocateDamage_AfterRefactoring(
    DamagableObject &io_object,
    const DamagableObject &i_base,
    unsigned int i_damage)
```

```
{
    AdjustHitpoints(io_object, i_damage);
    UpdatePower(io_object, i_base);
    UpdateVisualDamage(io_object, i_base);
    return(IsDestroyed(io_object));
}
```

Most important, other functions now have access to the separate functions that were originally performed internal to the damage allocation function. In particular, the destruction test could be useful in several places. Functions like this are generally the lowest level to which functionality is broken down for reuse, but there is a need for a higher-level structure for reuse to avoid bogging down large projects in a myriad of functions. One such language feature is the object provided by object-oriented languages discussed next.

Objects

Functional programming languages are useful for particular types of projects, but many modern applications benefit from what is known as *object-oriented* programming languages. These languages offer support for enforcing encapsulation and abstraction of data to varying degrees. Used properly, this allows a more intuitive mapping from the problem domain to source code, thus facilitating better organization for reuse. As with functions, one of the benefits of reuse is less duplication of code.

However, because objects generally hide data and implementation details, there is a danger that this could hinder some forms of code reuse. There are several steps to preventing this problem, first of which is to document the internal or private implementation of an object in addition to the externals. This documentation should be kept separate from the documentation of the public implementation to prevent confusion. The public documentation is used under normal circumstances, but if the desired functionality cannot be found, then the private documentation can be searched as well. If a solution is found within the private implementation, refactoring is generally required to make that functionality properly accessible. Do not fall into the temptation of cut-and-paste, as it will lead to the many problems we discussed earlier. Additionally, do not attempt to hack access to the functionality. Both of these are shortsighted methods that will lead to trouble later.

To make this concept clearer, let us look at an example. The code for the classes presented in this example can also be found on the companion CD-ROM in Source/Examples/Chapter2/object.cpp. Here is the public part of a terrain class that uses a hermite spline algorithm internally to interpolate the terrain height:

ON THE CD

```
class t_Terrain
{
public:

    /**     Construct a new terrain from a height map.
     *      @param     i_width - number of elements in
     *                     the height map along the
     *                     x dimension
     *      @param     i_length - number of elements in
     *                     the height map along the
     *                     z dimension
     *      @param     i_heightMap - array of height
     *                     values of size i_width *
     *                     i_length where the y value
     *                     of the terrain at (x,z)
     *                     is located in the array at
     *                     [(z * i_width) + x]. The
     *                     array is copied and is not
     *                     required after construction
     */
    t_Terrain(unsigned int i_width,
        unsigned int i_length,
        const double *i_heightMap);

    /** Clean up terrain resources.
     */
    ~t_Terrain();

    /** Get the height (y) value at (x,z).
     *      @param     i_x - value in the direction
     *                      of the terrain width
     *      @param     i_z - value in the direction
     *                      of the terrain length
     *      @return    Terrain height.
     *      @pre       0 <= i_x < i_width and
     *                 0 <= i_z < i_length where
     *                 i_width and i_length are the
     *                 values provided to the
     *                 terrain constructor.
     */
    double m_GetHeight(double i_x, double i_z) const;

};
```

From the public interface, it is not obvious and should not be obvious that the terrain class uses a hermite spline algorithm internally. The reason for this is straightforward: we do not want users of the class to rely on the implementation details. However, if we were looking for a hermite spline algorithm, it would be useful to see the internal implementation documented as well:

```cpp
class t_Terrain
{
public:

    t_Terrain(unsigned int i_width,
        unsigned int i_length,
        const double *i_heightMap) :
    m_width(i_width),
    m_length(i_length),
    m_heightMap(new double[m_width * m_length])
    {
        std::copy(i_heightMap,
            i_heightMap + (m_width * m_length),
            m_heightMap);
    }

    ~t_Terrain()
    {
        delete [] m_heightMap;
    }

    double m_GetHeight(double i_x, double i_z) const
    {
        int l_xIndex = static_cast<int>(i_x);
        int l_zIndex = static_cast<int>(i_z);
        double l_xDelta = i_x -
            static_cast<double>(l_xIndex);
        double l_zDelta = i_z -
            static_cast<double>(l_zIndex);
        return(m_HermiteSpline(
            l_xDelta,
            m_HermiteSpline(
                l_zDelta,
                m_GetHeight(l_xIndex - 1,
                    l_zIndex - 1),
                m_GetHeight(l_xIndex - 1,
                    l_zIndex),
```

```
                        m_GetHeight(l_xIndex - 1,
                            l_zIndex + 1),
                        m_GetHeight(l_xIndex - 1,
                            l_zIndex + 2)
                    ),
                    m_HermiteSpline(
                        l_zDelta,
                        m_GetHeight(l_xIndex,
                            l_zIndex - 1),
                        m_GetHeight(l_xIndex, l_zIndex),
                        m_GetHeight(l_xIndex,
                            l_zIndex + 1),
                        m_GetHeight(l_xIndex,
                            l_zIndex + 2)
                    ),
                    m_HermiteSpline(
                        l_zDelta,
                        m_GetHeight(l_xIndex + 1,
                            l_zIndex - 1),
                        m_GetHeight(l_xIndex + 1,
                            l_zIndex),
                        m_GetHeight(l_xIndex + 1,
                            l_zIndex + 1),
                        m_GetHeight(l_xIndex + 1,
                            l_zIndex + 2)
                    ),
                    m_HermiteSpline(
                        l_zDelta,
                        m_GetHeight(l_xIndex + 2,
                            l_zIndex - 1),
                        m_GetHeight(l_xIndex + 2,
                            l_zIndex),
                        m_GetHeight(l_xIndex + 2,
                            l_zIndex + 1),
                        m_GetHeight(l_xIndex + 2,
                            l_zIndex + 2)
                    )));
        }

    private:

        /**     Interpolate value using hermite spline
         *      algorithm and four values.
         *      @param   i_t - interpolate (i_t, i_h)
```

```
*                given (-1, i_s0) ->
*                    (0, i_h0) -> (1, i_h1) ->
*                    (2, i_s1)
*     @param     i_s0 - value at -1
*     @param     i_h0 - value at 0
*     @param     i_h1 - value at 1
*     @param     i_s1 - value at 2
*     @return Value at i_t.
*/
double m_HermiteSpline(double i_t,
    double i_s0, double i_h0,
    double i_h1, double i_s1) const
{
double l_t2 = i_t * i_t;
double l_t3 = l_t2 * i_t;
double l_r0 = i_h0 - i_s0;
double l_r1 = i_s1 - i_h1;
return(
    ((2*l_t3 - 3*l_t2 + 1)     * i_h0) +
    ((-2*l_t3 + 3*l_t2)        * i_h1) +
    ((l_t3 - 2*l_t2 + i_t)     * l_r0) +
    ((l_t3 - l_t2)             * l_r1)
    );
}


/**    Get the height (y) value at (x,z)
 *     grid coordinate.
 *     @param     i_x - value in the direction
 *                    of the terrain width
 *     @param     i_z - value in the direction
 *                    of the terrain length
 *     @return    Terrain height.
 *     @pre        0 <= i_x < i_width and
 *            0 <= i_z < i_length where
 *            i_width and i_length are the
 *            values provided to the
 *            terrain constructor.
 */
double m_GetHeight(int i_x, int i_z) const
{
    return(m_heightMap[
            (m_ClampIndex(i_z, k_LENGTH) *
            m_width) + m_ClampIndex(i_x,
            k_WIDTH)]);
```

```
            }

            /// Grid index clamp types.
            enum t_ClampType { k_WIDTH, k_LENGTH };

            /** Clamp an index based on width or length.
             *    @param    i_index - index to clamp
             *    @param    i_type - clamp to
             *                  width(k_WIDTH) or
             *                  length(k_LENGTH)
             *    @return    Clamped index.
             */
            int m_ClampIndex(int i_index,
                t_ClampType i_type) const
            {
                switch(i_type) {
                    case k_WIDTH:
                        if(i_index < 0) {
                            return(0);
                        } else if(i_index >=
                            static_cast<int>(m_width))
                        {
                            return(m_width - 1);
                        }
                        return(i_index);
                    case k_LENGTH:
                        if(i_index < 0) {
                            return(0);
                        } else if(i_index >=
                            static_cast<int>(m_length))
                        {
                            return(m_length - 1);
                        }
                        return(i_index);
                }
                return(0);
            }

            /// Terrain dimension.
            unsigned int m_width, m_length;

            /// Internal copy of height map.
            double *m_heightMap;

        };
```

To understand this, imagine someone else wrote the terrain class and now you are writing an animation class:

```
class t_AnimationChannel
{
public:

    /**    Construct a new animation channel from
     *     a set of keys.
     *        @param    i_frames - Number of frames in
     *                      the animation at one
     *                      per second.
     *        @param    i_keys - Animation value at
     *                      each frame.
     */
    t_AnimationChannel(unsigned int i_frames,
        const double *i_keys);

    /** Clean up animation channel resources.
     */
    ~t_AnimationChannel();

    /** Get the value at a particular time.
     *     @param    i_time - Time in seconds from
     *                      animation start.
     *     @return   Value at i_time.
     *     @pre      0 <= i_time < i_frames
     *                  where i_frames is the value
     *                  provided to the constructor.
     */
    double m_GetValue(double i_time) const;

};
```

Further, you decide to use a hermite spline algorithm to implement it. After searching the public interface documentation for your project, you find nothing. However, extending the search to private implementations, you come across the already existing piece of code. You can then extract it from the terrain class:

```
/**    Interpolate value using hermite spline
 *     algorithm and four values.
 *        @param    i_t - interpolate (i_t, i_h)
 *                      given (-1, i_s0) ->
 *                      (0, i_h0) -> (1, i_h1) ->
```

```
*                  (2, i_s1)
*    @param    i_s0 - value at -1
*    @param    i_h0 - value at 0
*    @param    i_h1 - value at 1
*    @param    i_s1 - value at 2
*    @return Value at i_t.
*/
double g_HermiteSpline(double i_t,
    double i_s0, double i_h0,
    double i_h1, double i_s1)
{
    double l_t2 = i_t * i_t;
    double l_t3 = l_t2 * i_t;
    double l_r0 = i_h0 - i_s0;
    double l_r1 = i_s1 - i_h1;
    return(
        ((2*l_t3 - 3*l_t2 + 1)      * i_h0) +
        ((-2*l_t3 + 3*l_t2)         * i_h1) +
        ((l_t3 - 2*l_t2 + i_t)      * l_r0) +
        ((l_t3 - l_t2)              * l_r1)
        );
}
```

After this and rewriting the terrain class to use the new publicly available algorithm, you can write your class:

```
class t_AnimationChannel
{
public:

    t_AnimationChannel(unsigned int i_frames,
        const double *i_keys) :
    m_frames(i_frames),
    m_keys(new double[i_frames])
    {
        std::copy(i_keys,
            i_keys + m_frames, m_keys);
    }

    ~t_AnimationChannel()
    {
        delete [] m_keys;
    }
```

```cpp
    double m_GetValue(double i_time) const
    {
        int l_timeIndex = static_cast<int>(i_time);
        double l_timeDelta = i_time -
            static_cast<double>(l_timeIndex);
        return(g_HermiteSpline(l_timeDelta,
            m_ClampIndex(l_timeIndex - 1),
            m_ClampIndex(l_timeIndex),
            m_ClampIndex(l_timeIndex + 1),
            m_ClampIndex(l_timeIndex + 2)));
    }

private:

    /** Clamp index to number of frames.
     *
     */
    int m_ClampIndex(int i_index) const
    {
        if(i_index < 0) {
            return(0);
        } else
        if(i_index >=
            static_cast<int>(m_frames)) {
            return(m_frames - 1);
        }
        return(i_index);
    }

    /// Number of frames.
    unsigned int m_frames;

    /// Values at each frame.
    double *m_keys;

};
```

Another step to assisting in reuse is based on the previous section on functions. Objects generally act upon their data through some form of functions. This means that the lessons about proper granularity and level of responsibility apply equally as well to the function associated with objects. The concept of limiting the responsibilities of a function to make its role clear applies equally to objects themselves. Clear object roles lead to a more reusable set of objects.

Templates

So far, everything we have been talking about has been language independent, but now we will take a moment to look at a language feature particular to C++: *templates*. Do not take this to mean that other languages, such as Eiffel, do not support similar features, just that we will be focusing on the C++ implementation in this section. If you do not know C++, some explanation will be provided to help, but you can skip to the next section if you desire. With that aside, let us look at how templates can reduce the duplication of code in addition to providing protection from errors.

First, imagine that C++ did not have templates. If you wanted to be able to sum an array of integers, an array of doubles, and an array of strings, you would need to write the following three functions:

```cpp
int accumulate_int(const int *i_integers,
    unsigned int i_size)
{
    int l_result = i_integers[0];
    for(unsigned int l_index = 1
        l_index < i_size; ++l_index) {
        l_result += i_integers[l_index];
    }
    return(l_result);
}

double accumulate_double(const double *i_numbers,
    unsigned int i_size)
{
    double l_result = i_numbers[0];
    for(unsigned int l_index = 1;
        l_index < i_size; ++l_index) {
        l_result += i_numbers[l_index];
    }
    return(l_result);
}

string accumulate_string(const string *i_strings,
    unsigned int i_size)
{
    string l_result = i_strings[0];
    for(unsigned int l_index = 1;
        l_index < i_size; ++l_index) {
        l_result += i_strings[l_index];
    }
```

```
        return(l_result);
    }
```

Additionally, if you later need to sum another type, then another function would be required. The major problem here is that each of these functions contains code with only minor differences. Each of these introduces potential for an error, and if the functions were even slightly more complex, an error would be extremely likely. Now here is the solution with templates:

```
template <typename t_ValueType> t_ValueType
    accumulate_generic(const t_ValueType *i_array,
        unsigned int i_size)
{
    t_ValueType l_result = i_array[0];
    for(unsigned int l_index = 1;
        l_index < i_size; ++l_index) {
        l_result += i_array[l_index];
    }
    return(l_result);
}
```

This not only encapsulates the functionality of the three separate functions into one instance of the code, it also accounts for future functions that are similar. Templates provide a method to write generic functions within the strict type system of C++. This supports the concept of generic programming, which is primarily aimed at reducing the duplication of code for common algorithms while maintaining type safety, thus allowing errors to be caught at compile time. Java handles this differently by requiring that all objects be derived from a single base class, and performs type checking at run time. Many modern languages support this concept of generic programming, but tend to do it in their own unique way.

For those who know C or C++, you might be thinking of other ways that this could be solved without using templates and without duplicating code. There are several solutions fitting this criterion, but templates offer two critical advantages: safety and clarity. Common errors can be caught at compile time with templates that other solutions would only be able to handle at run time. Since templates are an integral part of the C++ language, they are supported by editors and debuggers, which allows clearer and easier to maintain code to be written. To make this clear, here is such a solution:

```
#define accumulate_macro(o_result, i_array, i_size) \
    { \
```

```
    o_result = i_array[0]; \
    for(unsigned int l_index = 1; \
        l_index < i_size; ++l_index) { \
        o_result += i_array[l_index]; \
    } \
}
```

This solution suffers from several problems. It cannot return a value, so its use must be different from the previous functions. It does not show up as a function even though a call to it will look like a function call. Moreover, perhaps one of the most serious problems, debuggers treat it as a single line of code that makes it impossible to place a breakpoint, or debug stopping point, in the middle of the function. None of these problems exist for the template solution.

The point of this is to emphasize that although there might be several solutions to a problem that avoids code duplication, it is important to consider other factors such as compile time versus run time error checking before making a final decision. Knowing the advanced language features and when to use them is important to reducing code duplication; just be careful to fully comment the result to aid programmers who have less experience with the feature.

Generic Programming

C++ templates are an example of a language feature that supports generic programming. The language Eiffel also provides similar support for generic programming. Each of these allows a *type safe* approach to development that still provides the necessary tools to reduce code duplication.

To understand the importance of generic programming, you must understand the importance of type safety and how type safety relates to the development process. Type safety allows errors involving the use of invalid types to be caught at compile time rather than run time, which is when languages lacking type safety would be capable of catching these errors. In fact, in some languages that lack type safety, these errors would not be caught at all.

In either scenario lacking type safety, the errors occur at run time and this places the burden of finding them on the testing phase of development. This can be a serious disadvantage because the errors might not be caught at all. The major difference is that the compiler catches all errors related to compiling the source code, whereas testing requires the programmer to generate the proper test case in order to discover an error. If the test case is missed, the error will also be missed.

You should therefore always take advantage of language features that allow generic programming. Additionally, you should consider this language feature when deciding which language to use for your project. Support for generic programming can greatly reduce the risk of uncaught errors when used properly.

Preprocessor

What do you do if you are already using a language that lacks the proper functionality to help you prevent certain forms of code duplication? Not all hope is lost. Since most language source code is written in plain text, it is generally easy to apply text processing to the source code before passing it to the language compiler or interpreter. In fact, C and C++ programmers generally assume the existence of a standard preprocessor for handling certain forms of duplication that the compiler does not.

Before you get too excited, and start writing preprocessors to handle everything, you must also consider the disadvantages of them. The primary shortcoming of preprocessors is their lack of integration with the language, including tools such as compilers and debuggers. Even the standard C preprocessor often creates problems with most debuggers by making it more difficult to resolve the location of errors. Non-standard preprocessors also make it more difficult to share code and require more training for new programmers. What this means is that you should prefer solutions inherent in the language to those that use a preprocessor. However, since the solution is not always possible using the language, it becomes necessary to use preprocessors to avoid worse problems such as code duplication.

Earlier in the template section, you saw an example of the C++ processor at work when we presented an alternative solution not using templates. In this case, it was better to use templates because they are more integrated into the language. However, in the Java language there is currently no support for generic programming; all type checking and conversion must be accomplished at run time. In most cases, this reasonable approach makes programming in Java relatively easy to understand. Nevertheless, situations do occur when compile time type checking is necessary, for either optimization or for critical systems that cannot afford to throw an exception. In this case, writing a custom preprocessor is the only approach until the Java language is extended to support true generic programming.

One major disadvantage that often arises when writing custom preprocessors is their interaction with the compiler and debugger. Because the syntax of the code written for the custom preprocessor is not understood by the compiler or debugger, these tools can only generate information that refers to the files generated by the preprocessor. This problem can be reduced by providing a utility to translate locations in the generated code back to the location in the original code provided to the preprocessor. This can then be run by hand to locate errors and warnings in

---

PRACTICAL PREPROCESSING

One project found a situation when the C++ language, even with templates, did not support the proper syntax to avoid considerable duplication. The basic system was to evaluate a collection of conditionals representing rules, and execute the associated code if the conditional was true. The problem was that the number of rules was large, so a more optimal method than evaluating all conditionals on every pass was required. The solution to this was to provide object types for the conditionals that tracked their state and indicated when a change was made. By storing a list of conditionals an object was associated with, the conditionals could be reevaluated only if the object had changed. This provided a considerable performance improvement, but there was a problem.

The code for associating the conditionals with their corresponding objects was repetitive and syntactically cumbersome. It was easy to omit objects and make other minor errors. These errors were difficult to debug, and often did not appear except under particular conditions.

To solve this problem, a preprocessor was written that would take a file containing a list of rules made up of conditionals matched with code blocks. This preprocessor would parse each conditional to determine what objects were contained within it. Then, a new file was automatically generated containing all the necessary code for associating the objects and conditionals, plus matching the code block with the conditional to be run when the conditional evaluated to true. With this system in place, the code could be written with greatly reduced risk of error. As an added bonus, tracing and debugging information could be added to the automatically generated code to ease the debugging process. Without automation, this code would not have been written because of the tedium involved. In this case, the benefits outweighed the disadvantages and the preprocessor saved many errors.

---

the original code. The usefulness of this utility can be further increased if the compiler and debugger are extensible, allowing the utility to be directly integrated into these tools and removing any intermediate steps.

Aspect-Oriented Programming

A relatively new language feature, *aspect-oriented* programming, is emerging that can assist in reducing code duplication, particularly for debugging and profiling code. There are aspects of production code that will also benefit from the new language feature. You will likely start using it for debugging purposes, so you might wonder why worrying about code duplication in throwaway code is important. It is just as easy to create a bug in debugging code as it is in production code, but it can be even harder to track down that bug since you will be likely to look in the production code first. So, how can aspect-oriented programming reduce code duplication? Let us start with a little explanation of exactly what it is, and then present an example.

Aspect-oriented programming encapsulates development concerns that cut across the normal structure of the programming language being used. In other words, aspects are meant to encapsulate common code that does not easily fit within the normal structure of the language. In the case of object-oriented languages, this means that the common code aspect must be inserted, or woven, into multiple places across several objects. These insertions are similar in format, with only minor changes for each new location. Because of this similarity, it is common to use cut-and-paste techniques to perform these insertions that lead to code that is difficult if not impossible to maintain. Aspect-oriented programming, however, removes this code into a single location for clarity and maintainability, and the aspect code is automatically woven back into the object code at compile time (Figure 2.2A, B, and C).

This encapsulation therefore provides two major advantages: separation of the concern into an entity that can be maintained on its own, and the removal of code duplication caused by the concern's extent. Thus, aspects provide a mechanism for reducing cut-and-paste that is not present in the target language.

Defining aspects is generally accomplished by providing a method of describing where and how in the normal program structure to insert the aspect code. By abstracting the nature of these descriptions, aspects can be reused across multiple projects. The abstraction also reduces the maintenance of the aspect, requiring no updates to the aspect when only minor changes are made to the code into which the aspect is woven.

For example, calls to trace programming execution are often spread across multiple functions and objects within an application. The goal of aspect-oriented software development is to allow the trace concern to be encapsulated in an aspect for easy updating and removal from the final production code. There is one minor disadvantage to this scheme in that it separates part of the functionality of a function or object from the unit itself. However, in the case of crosscutting concerns, the bond between the aspect and the object is generally not as strong as the bond with the rest of the aspect code.

Now, to make this more tangible, let us take a class that must have an initialization method called before any other methods. The code for this example can also

A

```
Account
...
void calculateBalance () {
System.out.println ("void calculateBalance ()");
...
System.out.println ("end calculateBalance ");
}
...

Customer
...
void deposit (float amount) {
System.out.println ("void calculateBalance (" +
"float amount = " + amount + ")");
...
System.out.println ("end calculateBalance ");
}
...
```

B

```
aspect Trace
...
around( ... ) {
System.out.println (trace info );
proceed(...);
System.out.println (trace info );
}
...

Account
...
void calculateBalance () {
...
}
...

Customer
...
void deposit (float amount) {
...
}
...
```

C

```
Account
...
void calculateBalance () {
System.out.println ("void calculateBalance ()");
...
System.out.println ("end calculateBalance ");
}
...

Customer
...
void deposit (float amount) {
System.out.println ("void calculateBalance (" +
"float amount = " + amount + ")");
...
System.out.println ("end calculateBalance ");
}
...
```

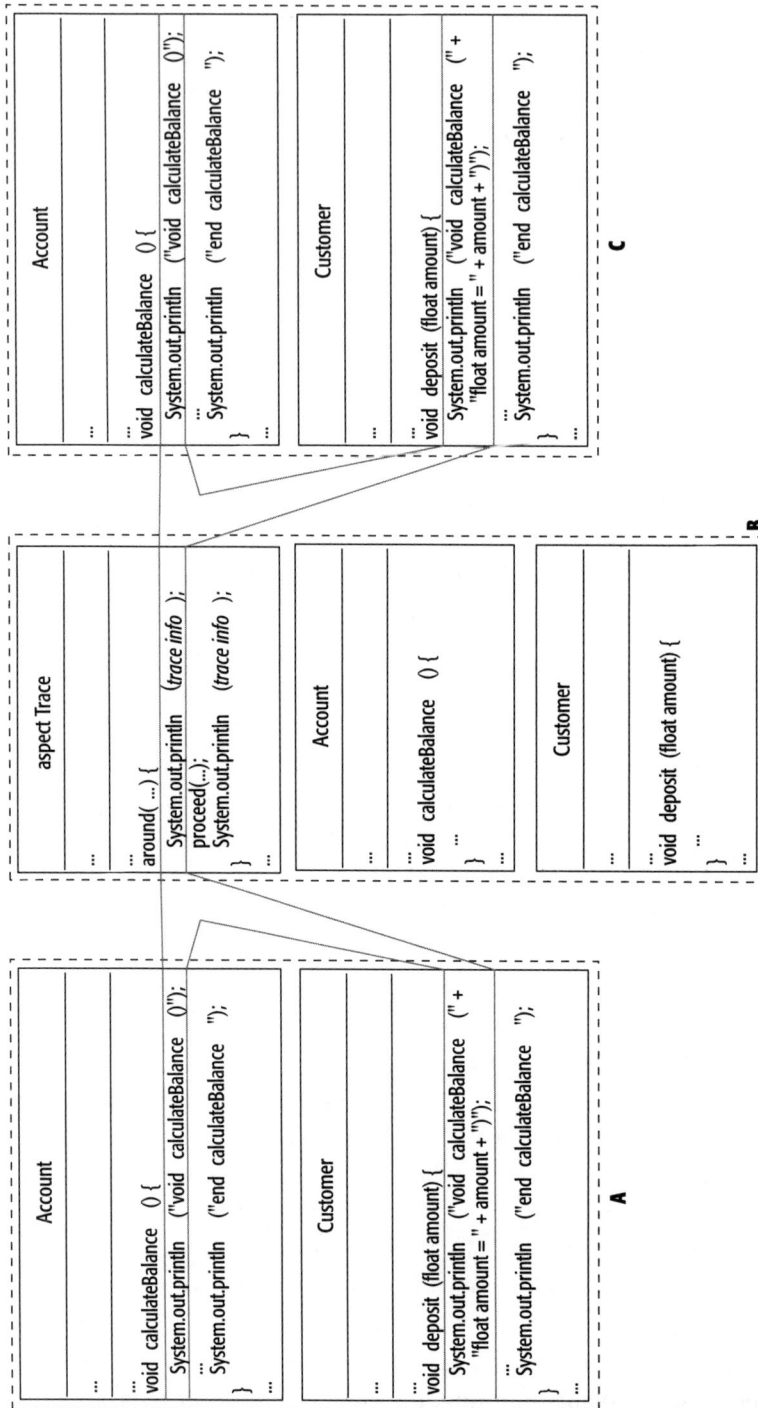

FIGURE 2.2 A) Code with tracing before using aspects, the programmer would have to edit each instance of the tracing code. B) Tracing code has been removed and collected into a single aspect; this is now what the programmer will view and edit. C) Tracing code is woven back into the object code; this would not normally be seen by the programmer.

be found on the companion CD-ROM in Source/JavaExamples/com/crm/ppt/
examples/chapter2/WithoutAspectExObject.java, Source/JavaExamples/com/crm/
ppt/examples/chapter2/AspectExampleObject.java, and Source/JavaExamples/com/
crm/ppt/examples/chapter2/AspectExampleAspect.java. Although it is preferable
to avoid this type of object, sometimes it is necessary. Here is a first attempt at this
class:

```java
/**
 * Example object that requires initialization.
 */
public final class AspectExampleObject {

/**     True when object is initialized
 *      and hence usable.
 */
boolean isInitialized;

/** Creates a new instance */
public AspectExampleObject() {
}

/** Initialize object. */
void init() {
      isInitialized = true;
}

/**     Do something that requires the object
 *      be initialized.
 */
void doSomething() {
    if(isInitialized) {
        System.out.println(
        "AspectExampleObject doing something...");
    } else {
        System.out.println(
        "AspectExampleObject tried to do");
        System.out.println(
        "something, but failed catastrophically");
        System.out.println(
        "because it was not initialized.");
    }
}
```

```
/**     More methods that require an
 *      initialized object... */

/** Uninitialize object. */
void done() {
    isInitialized = false;
}


}
```

The major problem with this object is the possibility that its methods can be called before it is initialized. This could cause all kinds of unpredictable results and mayhem. Therefore, to protect against that, we can add a test at the beginning of each method that throws an exception if the object is not initialized:

```
/**
 * Throw exception on uninitialized objects
 * upon method call.
 */
public final class AspectExampleObject {

/**     ... */

/**     Do something that requires the object
 *      be initialized.
 */
void doSomething() {
    if(!isInitialized) {
        throw new java.lang.IllegalStateException();
    }
    if(isInitialized) {
        System.out.println(
        "AspectExampleObject doing something...");
    } else {
        System.out.println(
        "AspectExampleObject tried to do");
        System.out.println(
        "something, but failed catastrophically");
        System.out.println(
        "because it was not initialized.");
    }
}

/**     More methods that require an
```

```
*    initialized object... */

/** Uninitialize object. */
void done() {
    if(!isInitialized) {
        throw new java.lang.IllegalStateException();
    }
    isInitialized = false;
}


}
```

Notice that the code had to be duplicated in the majority of the object's methods. This is the type of code duplication to avoid. What we really want is a way to place a single piece of code at the beginning of the functions without duplicating it. This can be accomplished succinctly with the following aspect:

```
/**
 * Throw exception on uninitialized
 * objects upon method call.
 */
aspect AspectExampleAspect {

    /**
     *    All methods on AspectExampleObject
     *    that cannot be called until
     *    the object is initialized.
     */
    pointcut
    afterInitMethods(AspectExampleObject aeo):
        target(aeo) &&
        call(* AspectExampleObject.*(..)) &&
        !call(void AspectExampleObject.init());

    /**
     *    Throw an exection at afterInitMethods
     *    pointcuts if the
     *    object is not initialized.
     */
    before(AspectExampleObject aeo) :
        afterInitMethods(aeo) {
        if(!aeo.isInitialized) {
                throw new
                java.lang.IllegalStateException(
```

```
                    "AspectExampleObject uninitialized ["
                    + thisJoinPoint + "]");
            }
        }

    }
```

Another advantage to this particular aspect is that it could be reused with other objects that follow the necessary conventions. This further reduces the unwanted duplication of code. Current support for aspect-oriented programming is limited, and even for the more complete systems like AspectJ it is still in the preprocessor stage. As this and other advanced language features become integrated into the core languages, it would be wise to take advantage of the new capabilities, such as reduction of code duplication, which they provide.

Automation

Sometimes it is necessary for data to exist in more than one location at one time. It is essential to remove human participation from this process. To accomplish this, you can use one of the many tools available for this purpose. One such tool is the code preprocessor that we mentioned earlier. However, code is not the only part of development that is prone to cut-and-paste problems.

Almost all modern applications have a number of files, or assets, that are not created from compiling code. As the number of assets grows larger, it becomes more difficult to manage. If information requires updating in more than one location, it is almost guaranteed that errors will occur. Let us examine some general tools that can be used to automatically duplicate information when necessary and discuss how programmers should use them.

DANGERS OF MANUAL DUPLICATION

One project team discovered firsthand the dangers of requiring information to be updated in multiple locations. In this case, the user interface for the application under construction used a large number of images that were to be placed in different locations. Unfortunately, the user interface configuration files required that the size of a control be provided. This meant that every time an image was resized, the corresponding control configuration also required an update. This led to many problems, one particularly problematic because it initially looked like the art had not been updated. After recreating the art three times, the real error was finally found, but not until after several team members wasted considerable time.

The most flexible and therefore most demanding tool at a programmer's disposal is the variety of scripting languages created for the purpose of file and text manipulation. These allow for quick implementation of tools that can handle a wide variety of tasks associated with copying and modifying assets. The primary disadvantage to using these tools comes when other programmers need to modify the tools. Depending on the scripting language chosen, other programmers might not understand the language. Even if they do understand the language, some scripting languages such as Perl have such limited structure that it can still be hard to understand what the script is doing. Perl can be excellent for quick one-run scripts that are then deleted, but sticking to a standard language with a reasonable amount of structure, such as Python or Ruby, is the best bet for any script that will be around for more than an hour. See Table 2.1 for a comparison of Perl, Python and Ruby. Another disadvantage of scripting languages is that they are generally interpreted, which requires that the interpreter be installed on the machine on which they are run. This second disadvantage is shared with most other tool solutions, so it should not be a major concern. If possible, keep a distributable copy of the files necessary for installation with the scripts for easy distribution.

TABLE 2.1 Comparison of Three Major Scripting Languages

	Perl	*Python*	*Ruby*
Code Readability	Poor	Good	Good
Community Support	Excellent	Good	Good
Object Oriented / Procedural	Both	Both	OO
Inheritance	Multiple	Multiple	Single / Multiple Mix-ins
Garbage Collection	Mark / Sweep	Reference Count	Reference Count
Regular Expressions	Integral	Library	Integral

Another obvious choice is to write your own automation tools in the same language as the application. This also has its advantages and disadvantages. The primary advantage is that the tools and understanding will already be there for members of the same team. Nevertheless, many of the higher-level languages used for application development do not have full-featured text and file manipulation utilities that come with the standard language. This can be alleviated to a degree by using third-party libraries to fill in this functionality, but you are then back to the necessity of ensuring that these are installed and understood by any programmer who must maintain the tools. An advantage over interpreted scripting tools is retained, however, because the tool users are not required to install anything additional. It can be argued that it takes less time to develop the tools in the interpreted scripting languages, but this is really only true if the programmer has a good grasp of the scripting language. In the end, the decision between scripted or compiled tools has to be made on a team-by-team and even tool-by-tool basis.

The final step in proper automation is to collect all the various tools, scripts, and batch files, and integrate them properly into the workflow of the programmer. A minimal and complete set of common operations that can be performed with one mouse click or a single command should be created. This usually requires a master script or batch file for each operation that sets off all required tools and scripts. If there are dependencies inherent in the process, it might be better to use common tools such as Make or Ant. These allow operations to be easily written to depend on other operations. Finally, it is beneficial to make these available from a single graphical user interface (GUI) or the IDE (integrated development environment) used by the programmer or programmers.

Avoiding Asset Duplication

While circumstances can dictate the use of automation tools, a better approach is to avoid the need for them altogether. If you have enough control over how the application reads the assets, it should be possible to avoid the need to read the same values from multiple locations.

This is best introduced by giving a simple example. Here is some pseudo-code that illustrates what could happen:

```
Laser
{
    integer range;

    initialize()
    {
        // ...
        read range;
        // ...
    }

    fire()
    {
        // ...
```

```
        use range;
        // ...
    }
}

AI
{
    integer range;

    initialize()
    {
        // ...
        read weapon range;
        // ...
    }

    check_target()
    {
        if target in range
            fire weapon;
    }
}
```

This requires the range value of the weapon to be available from two different locations; this puts undue responsibility on asset creation to synchronize these values. We can refactor this solution into the following pseudo-code:

```
Weapon
{
    integer range;

    initialize()
    {
        //...
        read range;
        //...
    }

    fire();
}

Laser derived from Weapon
{
    fire()
```

```
    {
        //...
        use range;
        //...
    }
}

AI
{
    check_target()
    {
        if target is in Weapon.range
            fire Weapon;
    }
}
```

Now there is only one location where the range value is read, thus avoiding duplication of the value in the assets. This is as important or more important than avoiding code duplication, because finding duplication errors in assets is often a more difficult task than finding code duplication. This becomes more likely in a large system. Therefore, it is a good idea for at least one person to understand the asset layout so that he can track down these duplications.

Generative Programming

Up until now, we have talked about practical methods to help prevent cut-and-paste that are available today. Even aspect-oriented programming, while innovative technology, is available for use as a solid working language feature for the Java language. However, other technologies are on the horizon that will be important to reducing code duplication, and even more important, information duplication in general.

To understand the importance of these technologies, first you must understand more about information duplication versus code duplication. Understanding that cut-and-pasting of code is error prone and detrimental to software development is only part of the struggle, because in general the code does not contain all the information that is necessary to develop a software product. In addition, there is information such as customer requirements, domain knowledge, and higher-level design and architecture. Much of the code written represents a translation of this information into a form the computer can read and execute. While it cannot be denied that some of this translation requires the creative skills of a human programmer, there are still many automated translations not being performed that would show significant gains in development time without corresponding loss in performance or completeness.

The topic of *generative* programming is much more complex and involved that a few pages could do justice, so here we will only look at how this future technology can reduce information duplication on a larger scale than current programming technologies. More information can be found in [Czarnecki00] and new research papers that are appearing continuously.

While automated programming has been a dream since near the beginning of programming itself, generative programming is an attempt to take a more practical approach to the idea. Rather than attempt to fully automate the process of software development, generative programming aims for the more realistic goal of automating only the processes that make sense to automate with the current technology level.

The primary idea is to take a family of related systems in a specific domain and automate the creation of applications from a high-level customer specification of the desired application features. Every piece of code is created, integrated, and built automatically except for custom requests that are not available as part of the current generators. These custom components could even be integrated into the generation process for future use, improving the automation process even further. This frees the developers to work on only those parts of the application that cannot be automated, greatly increasing development time.

There is, however, an obvious initial cost to building the necessary libraries and generators that can compose these applications. Here some of the language features and technologies we discussed earlier come into play, along with a few other experimental technologies. To facilitate the generation of code to support a range of features, the concept of active libraries [Czarnecki00] is necessary. An active library acts as a compile-time code generator parameterized on the desired features for the specific instantiation of the library. This provides the necessary flexibility for generating a range of applications without the large run time overhead that standard libraries impose. Language features such as templates and aspects that support techniques such as generic programming and aspect oriented programming are essential to creating these active libraries.

While many of the necessary language features already exist, their use is hindered by the difficult syntax and layout capabilities of many current editors. One possible resolution to this problem is the current effort by IntentSoft to develop the *Intentional Programming* environment. Starting as a research project at Microsoft, the main concept of Intentional Programming is to represent the software as a construct that preserves the intentions of the programmer rather than forcing the programmer to translate those intentions into text. This is accomplished through an extensible environment of editors, compilers, debuggers, and other programming tools. This allows the concepts necessary for developing active libraries to be generated, edited, and viewed in a more intuitive fashion that greatly assists the development process. Additionally, Intentional Programming supports another goal of generative programming by allowing the source to be represent and stored using domain specific terms and concepts. This allows the library developer to provide a structure that is directly related to the parameterization of the library based on domain-specific knowledge (Figure 2.3). This technology, and ones similar to it, should be kept in mind for adoption on future projects, particularly if reuse is important.

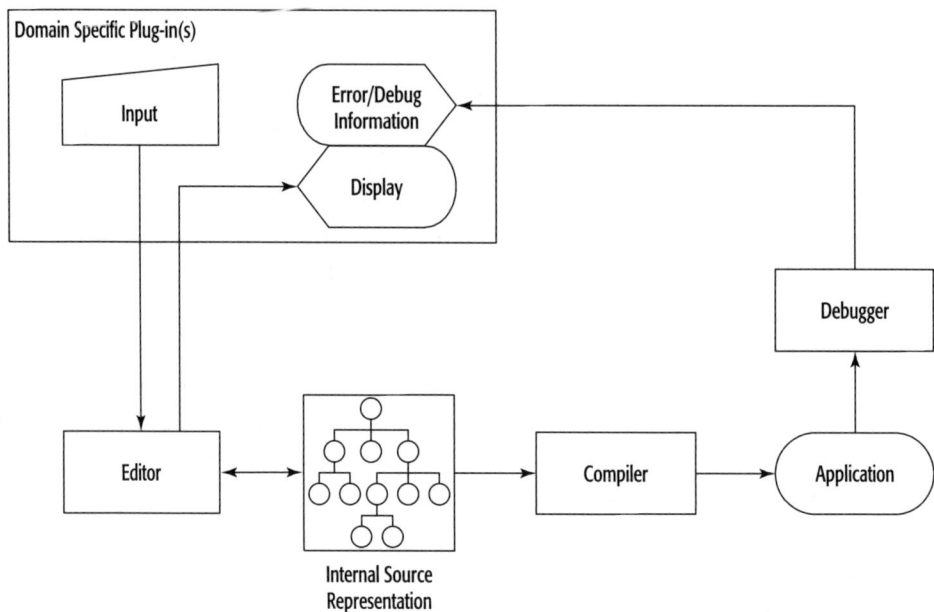

FIGURE 2.3 The editor, compiler, and debugger work together with one or more domain specific plug-ins to allow the programmer to view and interact with the source in a domain-specific manner.

So far, we have been talking about what generative programming is, rather than how it reduces information duplication. Obviously, since it relies on several of the language features that reduce code duplication, it is meant to reduce code duplication. With the introduction of Intentional Programming, however, we see the first glimpse of its uses in reducing information duplication. By allowing the programming environment to be extended by domain-specific concepts, it can also be extended to support design knowledge as well. Thus, Intentional Programming provides the necessary technology to represent the domain knowledge and design information directly in the source of the application or library.

Generative programming aims to go one step further than this, taking customer specifications at a high-level, translating those into the correct parameters with which to instantiate the necessary active libraries, and then hooking everything together to form a complete application. Small slots are left for developers to implement code that is not available in the active libraries, and therefore must be custom coded based on the customer's individual request. Thus, code duplication is reduced through the use of active libraries and their associated language features and programming techniques. In addition, information duplication is reduced by incorporating domain and design knowledge into the source through systems such as Intentional Programming. Finally, information duplication is further reduced by creating a generative system that requests only a high-level feature description and a small amount of custom code. Human-editable information is reduced to the minimum necessary, reducing the possibility of human error and therefore project risk (Figure 2.4).

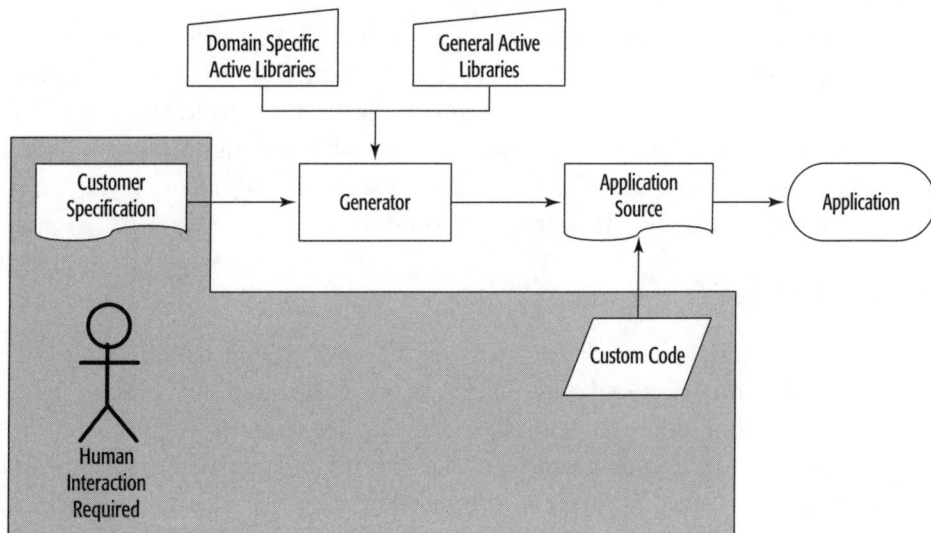

FIGURE 2.4 Generative programming only requires human interaction during the customer's feature specification and the bits of custom code that are not available in the active libraries.

Note that generative programming as a whole is not for every project, but if you are creating applications for multiple customers that are similar in nature but with several variation points, then generative programming should be of great interest to you. Even if your applications tend to differ greatly, you can still benefit from the active library concepts that are a necessary foundation for generative programming. Active libraries are particularly important to overcoming the performance concerns that plague NIH Syndrome, another of the major illnesses.

One caveat does arise when adopting the different technologies of generative programming such as active libraries. Many of these technologies require more processing power and can therefore increase compile times. This results from both their inherent complexity and the newness of the technology. As time passes, the speed of editors, compilers, and debuggers will increase and remove part of this overhead. The rest must be removed by increasing the power of the hardware used for development. Keep this in mind when deciding how to use these new technologies, as lengthening the compile times of code that must be compiled for the many changes and programmers on a project can add substantial time to development.

CURE

Even if the preventative measures are followed, you will encounter instances of cut-and-paste that requires reworking. Mistakes can be made, or sometimes it is not worth it initially to rework the application to avoid the cut-and-paste. Of course, you will also work with code from other programmers who did not follow the proper guidelines. Before we look at how to apply a cure to existing code, we must decide when it is appropriate to do so.

When to Refactor

Refactoring is a balancing act between the work expended refactoring and the work saved by the refactored code. There can be no hard-and-fast rules that always apply, but you can follow guidelines as you learn for yourself when refactoring actually improves development time. To take a better look at when to refactor in the context of cut-and-paste and code duplication, we will start with the most straightforward case and work our way up from there.

The simplest case occurs when you create the duplicated code yourself. For instance, you are adding functionality to one module that is available in another module. Sharing the code would require a reasonably large amount of reworking.

Instead, you cut-and-paste the code and modify it to fit the new situation. This is an acceptable solution when avoiding the code duplication would require considerably more effort than you expect to save from cleaner and more robust code. Be sure that there is a large difference between the work performed versus the work saved, and do not trust only in the accuracy of your estimations.

What happens when you suddenly realize that there is another instance where you require similar code? Refactoring should definitely occur here; otherwise, you will find yourself forgetting about one of the cases. Generally, two copies are acceptable but three should not occur, and four is almost guaranteed to cause problems. In addition, if you do have two copies of similar code, make sure the comments indicate the location of the other code and provide a warning about the effects of modifying or copying the code without taking into account the other copy. It is essential that both copies be commented in this manner, as it is not possible to predict which copy might be modified by you or another programmer.

Now we move on to the more difficult topic of refactoring legacy code, which generally comes from a past project. We will assume that the code is currently working; otherwise, refactoring would be required regardless and you should consider scrapping the old code entirely and writing your own. At this point, you might be thinking, if it's not broke, why fix it? There are several reasons to consider refactoring legacy code. Start by considering what to do when you discover that code you modified also exists in another location. Chances are you want to make the changes in both locations. This is a good time to go looking for more copies, particularly if the ones you did find did not comment on the duplication. This is an indicator of poor coding practices, and it would not be surprising to find multiple cut-and-paste copies.

Because duplicate code can cause considerable problems and be difficult to track down, it is a good idea to consider refactoring an entire module if you expect to make major changes to that module. This also can lead to a better understanding of what the module does and how.

Refactoring Tools

One of the reasons why many programmers avoid refactoring is the tedium of making the necessary modifications, which can sometimes involve sweeping changes that cover large sections of the code base. Without the necessary refactoring work, cut-and-paste code is easy to create and can have long-lasting effects during development. The obvious choice is to make refactoring easier. One question you might ask yourself is, "what tools can help reduce the time it takes to refactor the code?"

The optimal tool for refactoring must understand the language that is being refactored, and provide support for common refactoring operations. An example of this, and one of the first refactoring tools available, is the refactoring browser for the Smalltalk language. The refactoring browser allows refactoring to be achieved with minimal work from the programmer by automatically taking care of aspects such as renaming and temporary variables.

Other languages generally do not offer such strong support for refactoring tools, but this will change over time. Java is perhaps the only other language to have robust refactoring tools available, among them the IDEA integrated development environment from IntelliJ. IDEA supports many standard refactoring operations in the Java language, removing the need for tedious and error-prone manual refactoring.

A couple of examples should provide a better understanding of the true advantages provided by refactoring tools in IDEA. We will start with a simple renaming example for the following class:

```
public class RenameField
{
    boolean fieldToRename;

    // ...

    RenameField(boolean renamedField)
    {
        fieldToRename = renamedField;
    }

    // ...

}
```

Now we want to change the name of `fieldToRename` to `renamedField`. If we do a straightforward search and replace, we obtain:

```
public class RenameField
{
    boolean renamedField;

    // ...

    RenameField(boolean renamedField)
```

```
    {
        renamedField = renamedField;
    }

    // ...

}
```

Notice the assignment `renamedField = renamedField`, which should be `this.renamedField = renamedField`. However, by invoking the Rename refactoring in the IDEA interface, it correctly handles the change, resulting in:

```
public class RenameField
{
    boolean renamedField;

    // ...

    RenameField(boolean renamedField)
    {
        this.renamedField = renamedField;
    }

    // ...

}
```

Now let us look at a slightly more complex example. We start with the following class:

```
class ExtractMethod
{
    int length;

    int width;

    // ...

    boolean isLargerThan(int height, int volume)
    {
        return(volume < length * width * height);
    }
```

```
// ...

}
```

Now imagine we want to use the volume calculation for another function. Since we do not want to duplicate the code involved, we select `length * width * height` and apply the Extract Method refactoring option to that. This results in the following functions with the only user input required being the name of the new function:

```
boolean isLargerThan(int height, int volume)
{
    return(volume < volume(height));
}

private int volume(int height) {
    return length * width * height;
}
```

Notice that it correctly creates the necessary parameter and replaces the original code with the correct function call. Although not shown in this example, it can also handle temporary variables correctly. Expect other editors to follow suit as the advantages of this become clear. Contact your favorite editor developer and encourage them to include this support as soon as possible. In the meantime, it is still better to do refactoring even if it must be done by hand. Let us look at some tools that can help with manual refactoring.

Other languages do not have as complete support for refactoring, but there are still tools available that can help. The first tool that is a necessity is a good multiple-file search utility, which is included in most modern integrated development environments (IDEs). While you might be tempted to look for a multiple-file search-and-replace utility, this is usually not a good idea as it encourages blind changes to the code that are prone to error. However, there are even better utilities for refactoring searches that more advanced IDEs possess. These allow context-sensitive searches to be done with knowledge of language constructs, allowing the differentiation of similar names based on such concepts as type and instance. The reason why these search utilities are of prime importance to refactoring lies in the fact that changes often cut across the boundaries of language constructs and therefore across files.

Beyond Refactoring Tools

While human involvement will still be necessary for a long time to come, it is possible to perform some refactoring to remove code duplication using an automated process. This automatic restructuring of code would provide considerable benefits to the development process by removing the time-consuming task of cut-and-paste refactoring. Keep an eye on research in this area and encourage your favorite editor company to pursue more work on refactoring tools and automated refactoring.

A taste of what is to come can be found by looking at *Guru*, a hierarchy-restructuring tool for the Self programming language. This tool was developed at the University of Manchester as part of the PhD research of Ivan Moore. This is an initial example of automatic refactoring of an object-oriented language. Expect automated refactoring to eventually become a standard part of future IDEs.

Importance of Testing Redux

When we talked about premature optimization, the importance of testing was stressed to achieve maximum optimization with minimal risk. The same treatment of testing applies when dealing with cut-and-paste. Fixing cut-and-paste code often involves major changes to the application code, but this should produce no changes in application functionality. A proper set of application functionality tests can minimize the risk that this will occur. These tests should be automated, and run after each individual change, as with any other form of refactoring. Throughout this book, the importance of testing will continue to be stressed. Although this might seem repetitive, testing is an extremely important and often overlooked aspect of minimizing development risk. If you are not already using at least a minimal set of tests, you would be well advised to consider how testing can be integrated into your development process.

Just as with the automation of cut-and-paste operations, the majority of tests should be automated as well. In this case, minimizing human involvement is not so much to prevent error, although that is a concern, but to make it less bothersome to perform the tests. Without this necessary step, many programmers will often skip the testing phase of coding and just pray that the code works. This can be disastrous, especially when refactoring several similar sections of code into one parameterized section of code. This type of refactoring is common when removing code duplication. Due to small differences between the copies of the code, the parameterization process can easily introduce a small error in one or more of the new parameterized calls that could go unnoticed until much later without proper testing.

RELATED ILLNESSES

The other illness that contributes most to the CAP Epidemic is Myopia. Cut-and-paste represents a quick-and-dirty short-term solution, but inevitably leads to problems in the long term. A greater focus on the long-term effects will reduce the amount of cut-and-past as the detrimental effects become obvious. Another contributor to code duplication is NIH Syndrome. Even if you are duplicating the functionality of code that is not developed in-house, the resulting effects of having two copies of the code can still rear its ugly head. Another programmer might decide to use the third-party algorithm, and then you will end up supporting two sets of code that do the same thing. The next chapter takes an in-depth look at this and other results of NIH Syndrome.

Using cut-and-paste when building the basis for the application leads to a brittle framework that is an occurrence of Brittle Bones. Changes become difficult, and as functionality is built on top of the framework, it uses different instances of the same functionality. Future changes to one instance but not the others can cause surprising results. Another illness that often manifests by accident when using cut-and-paste is i. This results from the placement of code into a new context without updating the variable names, which leads to poorly named variables that can mislead another programmer reading the code. A myopic solution to this is to ensure that you update the variable names, but a much better ultimate solution is to avoid the duplication of code altogether.

FIRST AID KIT

Ironically, the primary means of preventing a CAP Epidemic is not to use a particular tool: cut-and-paste. Nevertheless, there is more to it than that because if you still duplicate code and information without using cut-and-paste, you will be missing the point and creating extra work for yourself. You must therefore use tools that allow you to remove the need for duplicating information, particularly the need for programmers to perform the duplication manually.

You are likely to be using one of the primary tools for this purpose already, and that is the high-level language with which you are programming. The trick is to use the language to its full advantage, with the idea of removing duplication in mind. There are also language extensions, such as the aspect-oriented programming extension to Java named AspectJ, that provide additional methods for reducing duplication in ways that are not available in the standard language.

While the best method for preventing a CAP Epidemic is to avoid duplication entirely, it is not always possible due to limitations in the language or technology being used. This is where scripting languages such as Ruby, Python, Perl, and many others become important. Scripting languages allow fast implementation of useful utilities for automating information duplication tasks. Several of the common build utilities, such as Make and Ant, also have features that can assist in automating duplication.

If a CAP Epidemic has already reared its ugly head through your project's code, it is time to refactor that code to remove the redundancies. To this end, refactoring tools such as Smalltalk's refactoring browser and the refactoring options offered by IntelliJ's IDEA for Java are excellent utilities to remove the tedious repetitive chores of refactoring. If you read Chapter 1, you will notice that refactoring tools were also mentioned there. Refactoring tools are important in curing many programmer illnesses, but are particularly useful for Premature Optimization and the CAP Epidemic due to their status as major illnesses, and corresponding frequency of occurrence. Refactoring is an extremely useful process to begin with, and by removing the tedium and error-prone manual implementation, it becomes an essential part of a programmer's repertoire.

SUMMARY

The symptoms of a CAP Epidemic are:

- Bugs that keep reappearing in different code locations.
- The heavy use of search and replace for editing code.
- Nonsensical comments and names due to a change in context of the copied code.
- Strange bugs caused by forgetting small changes to copied code.
- Complex procedures required for updating a single piece of information.
- Duplicate occurrences of any human-editable information.

To prevent a CAP Epidemic:

- Know what project code is available for reuse.
- Know what external libraries are available for use in place of implementing the functionality internally.
- Use high-level languages and their features such as functions, objects, templates, preprocessors, generic programming techniques, and aspect-oriented programming techniques.
- Automate any unavoidable information duplication using scripting and other automation tools.
- Code to avoid the need for asset duplication.
- Keep up to date on new programming technologies such as Intentional Programming and Generative Programming.

To cure the effects of a CAP Epidemic:

- Use refactoring to parameterize duplicate code and subsequently move it to a single location.
- Take advantage of refactoring tools.
- Keep up to date on the advancements in automating refactoring.
- Test often to ensure that the refactoring does not change the behavior.

3 ┆ NIH Syndrome

DESCRIPTION

Many programmers are opposed to using code other than their own. However, while there are reasons to be wary, it does not mean that you should refuse to trust any third-party code. Unfortunately, this seems to be the most common mindset when it comes to using external code. If the code was Not-Invented-Here, which can mean anything from within the company to an individual level, then it cannot be good enough for the task at hand. Some of this comes from legitimately bad experiences, but most of the time it is more inherent in the way the programmer thinks. This type of thinking is extremely problematic for software development, and as the complexity of software increases it will only become more so. It is imperative that something be done to curb this trend of rewriting code for every project at every company.

Fortunately, there are ways to improve the use of third-party code that will ease the mind of the paranoid programmer. By adopting these practices, you both remove the psychological barriers to using other's code and reduce the real risks in doing so. You can also demonstrate to other participants how the risk is minimized to convince them that using the third-party code will not bring the project crashing down.

SYMPTOMS

Even more problematic than the illness is the fact many developers overlook that the problem exists, which does not allow preventing or curing NIH (Not-Invented-Here) Syndrome. To get around this, we first look at the symptoms that indicate the existence of the illness and why it is a problem. Only then can we move on to fixing the problem.

Control Fanatics by Nature

Programming represents a creative process that produces a deterministic result; in other words, the results of programming are usually predictable and therefore controllable. At least that is how it is perceived. Because of this, there is a tendency for the profession to attract those who desire more control over the product of their efforts. Programming offers a level of control that is not possible in the random world of nature. This tendency is enforced by the benefits a controlling personality brings to software development. Process and attention to detail are required to complete software projects on time and on budget.

However, perhaps a better description for many programmers is control fanatics. The very personality trait that helps in many respects can be very destructive in certain areas of programming. Obviously, one such disadvantage is embodied in NIH Syndrome. The desire to avoid using other programmers' code is at the core of an attempt to avoid losing control. Most programmers exhibit this trait to some degree, and will often take more work upon them in order to maintain control of the code. This is not to say that this is an entirely irrational behavior, but that there are better methods for handling the situation than avoidance. In order to understand how to change the thinking patterns related to NIH Syndrome, it is essential to understand why programmers are reticent to give up control of the code base.

What Are We Afraid Of?

Fear is a fundamental driving force of human nature, so it is not surprising that it manifests itself even in the logical world of software development. As with many other aspects of life, fear serves a useful purpose in avoiding disaster, but the cost is paid by the occurrence of misdirected fear. This type of fear is caused by a misunderstanding of the cause and effect of a situation, and it brings about actions that are in reality unneeded or even harmful to the desired goal. It is therefore important to distinguish fear from risk management, where fear is a condition brought on by a lack of information, and risk management is the analysis of information to determine the correct practices to avoid. With this in mind, let us look at some fears about using external code that do not fall under risk management. Later we will look closer at the valid points that these fears do raise, and how to properly manage the risk that the rational part of these fears represents.

Performance

One of the most prevalent reasons for avoiding third-party code is the concern over performance. Programmers are not likely to admit that code they did not write will perform adequately. However, before we even deal with this issue, a more important reason why performance should not be a concern is dealt with in the Premature Optimization illness. To summarize, performance should only be a concern at the end of development or under special circumstances before then, and optimizations should not be attempted before gathering profiling information to determine where optimizations are required. For a more in-depth discussion of why this is and how to prevent it, refer back to Chapter 1, "Premature Optimization." This principle should be applied to external code in addition to internal or personally written code, hence its relevance to NIH Syndrome.

The other important point to keep in mind when deciding whether to fear third-party code is that you cannot be an expert in every aspect of software development. Some third-party libraries and frameworks are developed by experts in their field making them more likely to be complete and efficient than a non-specialist programmer could possibly do for the same price. Take advantage of this when possible and concentrate on the areas where your experience is best used.

SOUND DECISIONS

One of the most overlooked disciplines in the interactive entertainment industry is sound programming. This is often given as a side project to any programmer with free time, even if they have little or no real experience with sound programming, and are unlikely to understand sound theory and design. This caused problems for one typical project that was targeting the PlayStation 2 console system. Several programmers worked on the sound system, but this task was only one among many that they had on their plate. Without the time or expertise to concentrate on the sound system, the project ended up with a simple yet still bug-ridden sound implementation as the project drew near completion. Although the bugs were removed, which took a considerable amount of time, the sound system ended up being simple and unimpressive.

Another project used a third-party sound implementation created by programmers dedicated to understanding and implementing advanced sound on the PlayStation 2. While it still took time to integrate the library, there were fewer bugs and more features. This not only saved development time, but it also gave the end user of the product a richer sound experience.

Learning

Programmers are often afraid that they will have to learn a new system and new concepts that will take time. Some do not want to do this, while others fear they will be severely side-tracked and stop being productive as they pursue only the learning experience. While these points are valid, avoiding the use of outside code so that you do not have to learn the interface is a manifestation of fear rather than proper risk management. To bring the issue back into the realm of risk management, we need to consider how to manage the learning experience to balance the need for using the new code with the time spent learning how to use the new code.

Some might still be reticent to learn new concepts, feeling that they can manage with their current set of knowledge. This is a disastrous line of thinking for a programmer. If you are one of the ones who want to stop learning, you might want to consider a career change. The world of computer science is in constant flux and the rate of discovery continues to accelerate. If you have trouble learning the new concepts at the rate necessary, it would be beneficial to get some training in how to study properly, but if you simply do not want to do the learning, then you will need to evaluate how you expect to compete in the industry.

Missing Features

Another common fear is that the third-party library will be missing a feature that you require later. Software development is an uncertain process that can easily spring a surprise on an unsuspecting programmer at any time during development. Therefore, it is certainly possible that necessary features will be missing from external code and libraries that were acquired earlier. However, the feared result of this is usually blown out of proportion to the actual impact. Most often ignored is that if this feature were not present when evaluating the third-party code, it would also not have been included in the regular programming schedule if the third-party code were not used. This means that there would still be schedule consequences to this oversight with or without the external libraries.

At this point, an obvious argument emerges, saying that it is more difficult to add a feature to other programmers' code and therefore it will take more time anyway. This is certainly true, but the impact can be offset by the time saved in not having to develop and test the remainder of the external code. Thus, it is really a matter of evaluating the potential risks and the corresponding reward potential rather than taking a single risk out of context. In addition, we will later look at some techniques to improve the speed with which fixes and features can be taken care of for third-party libraries.

Debugging

It is hard enough to track down your own bugs, so it is natural to fear tracking down bugs from other programmers. What if you use a third-party library and it has an error in the code? Since you didn't write it, the error will be hard to find. The time spent looking for it will be wasted when you could have been doing useful work. Moreover, what if you don't have the source code, how will it get fixed at all? Maybe you should just forget about using external code altogether.

Now that we have seen the path fear can lead us down, let us consider what the tradeoffs are and then decide. First, if you must develop functionality yourself there is a guaranteed overhead versus using existing code. How much this makes a difference depends entirely on the code and its complexity. Next, and more important, when you write the code yourself, it will have bugs as well and since it is in-house will have a smaller test base for working those bugs out. Suppose you had to spend two weeks learning a third-party library and another week tracking down a bug in the library, for a total of three weeks. That might seem like a lot of time, but then consider that to do it yourself you spend a month writing it, and another two months testing and debugging it. This is a common example of the real tradeoffs involved. As with all decisions, it is dependent on context, but the odds favor already written code. Later, we will talk about how to improve your chances with proper methods for choosing and integrating external code that will reduce time spent debugging.

Deadlines

One of the primary reasons why we worry about learning a new system and debugging it if something goes wrong is missed deadlines. Except for a small percentage of the industry, software development is primarily a commercially driven enterprise. This means schedules to meet and milestones to complete. Perhaps in the past, you have experienced schedule delays caused by the use of external code, or missed a milestone because of a third-party library. Even if you have not experienced this personally, you are likely to work with someone who has suffered from this very situation.

However, it is a mistaken notion to solely blame the use of external code and libraries for these problems. A common scenario that can cause problems goes as follows. You approach the project manager and explain that a set of sound libraries is available from a commercial sound software developer. Since your team needs that functionality, you ask if you can purchase the library for use on your project. The manager gets excited, approves your purchase, and sends you off with the necessary paperwork. He then looks at the project task list and removes the sound implementation task, since it is already written by a third party. Suddenly, you are expecting to finish everything a month sooner. Unfortunately, it takes two weeks to get the library integrated with the application and bug free. This causes the next milestone to be missed. The lesson to take away from this is not that using third-party code is bad; instead, realize that it is not a panacea and requires proper scheduling and risk management. Treated properly, external code is no more likely to cause missed deadlines than internal code.

What Should We Be Afraid Of?

On the other side of the coin, there are things to fear if you do not take advantage of available code and libraries. The symptoms that you will inevitably see are what you should really be afraid of because they guarantee the loss of time and money on a project.

Duplication of Code

If a solution already exists, and you implement it again yourself, you are guaranteed to be creating similar code for the same task. This is a waste of resources unless there is some substantial benefit the new implementation has over the one that was already written. As we just discussed, many of the common reasons why people believe the new implementation will be better are not as concrete as they first appear. What you are usually left with is lost programming time and an implementation that could have been integrated much earlier. Additional time can easily be lost waiting for the new implementation to be completed and tested.

Worse yet, if use of third-party libraries and code is not coordinated properly, some programmers on a team might use one implementation while others write their own. This leads to maintenance problems and other errors that are common to the Cut-and-Paste (CAP) Epidemic. To avoid this, use of libraries and external code should be openly encouraged so that it can be maintained and coordinated. Fear of third-party code will only lead to poor communication and hence a corresponding loss of control over code duplication.

Lists represent a very common algorithm that is duplicated by multiple programmers on the same project. Occasionally, projects even go so far as to have an implementation for each individual list. Many third-party libraries also have their own lists. Applications can end up mixing custom lists with several different third-party lists. Look through your application and see if this proliferation of list implementations is occurring. Now think about how much developer time was spent learning, implementing, and maintaining each of these lists. If instead, the project had settled on one implementation, such as the Standard Template Library for C++ or the Collection classes for Java, this time could have been better spent elsewhere and the risk of error would also have been reduced.

Duplication of Work

Duplication of code implies duplication of work and effort, but that is not the only source of duplicated work when all third-party solutions are rejected. There is also a duplication of testing and debugging effort involved in getting it working. This is handled to one degree or another by the original developer of the code. In addition, if the library or code is popular, there might already be a wide base of users. They are implicitly doing their own testing and debugging, saving you time and effort.

There is also the research time required to choose the proper data structure and algorithms. This is often overlooked, even during internal development. Chances are, if you are working on something internal, the time to research it has not been scheduled. This leads to quick and often inappropriate solutions that require revision later. On the other hand, a middleware developer making products for other developers has a real interest in spending time to make their code use the best algorithms and data structures for the job. Even if a mistake or two is made, these are often corrected if you are not the first customer to use their product. As you can see, there is a lot more time and resource considerations involved than the simple writing of code.

Is Third-Party Code Better?

The era of the Renaissance man is long gone and unlikely to reemerge soon. No longer can one person be knowledgeable in all areas of science, or even a subset of the knowledge such as computer programming. In order to create the structures of the modern world, real or virtual, it takes a group of people who each have their own specialty that they understand in detail. Therefore, it is likely that you will not be able to provide all the expertise your project requires.

Third-party libraries and code can provide a vehicle for this expertise. Developers of code meant for use by other developers are likely to seek out the most experienced programmers in the field in which their code will be used. Because they generally have a narrower focus, this knowledge is maintained and improved at a rate not possible with the distractions of other programming tasks to worry about.

PREVENTION

When we talk about preventing NIH Syndrome, there are two main goals. The first goal is to eliminate the automatic reaction to rejecting externally developed code and libraries. The second goal, which is essential to achieving the first, is to reduce the risks of using third-party code and libraries by adopting methods for handling integration and problem fixes smoothly. Once these goals are achieved, you can save development time by taking advantage of work that has already been done.

What Are Middleware and Open Source?

The best place to start is by describing what we are encouraging you to use. So far, we have been using the terms *third-party libraries* or *external code* to describe what those suffering from NIH Syndrome are trying to avoid. This was sufficient for discussing the symptoms, but to understand how to prevent decisions being made while suffering from NIH Syndrome you need to understand what decisions are affected. Boiled down to the core, we are discussing the adoption of any code written by a programmer other than yourself for use in your own code. There are, however, many forms this code can take. Each form requires slightly different handling and consideration.

The first differentiation is a common and important one: money. One choice is to go with a commercial solution often referred to as *middleware*, and to pay to use other developer's code. The other choice is to use free code, be it open source available to everyone or code written by someone else within the same company that is available to members of the company freely. The cost of the product does not necessarily indicate the quality, but there is a distinguishable difference between the best commercial middleware and some college student's freely available side project. The other big difference between commercial and free code, which we will talk about next, is the level of support that is given.

The next distinction of note is whether the source code is fully available or just the interface along with binary libraries. This has a large impact on the way the external code is integrated and maintained within your project. We will talk more about decision-making research into these and other differences shortly, but it is obvious that having the source code available offers more flexibility in integration and maintenance. Although less important, there is also a distinction between low-level libraries, high-level APIs, and application frameworks. We will go into detail about each of these as we progress through our discussion.

Technical Support

Problems will happen, but if you are properly prepared, they can be handled smoothly. This makes the level of support for third-party code an important consideration when there is a choice between different sources. There are many levels to technical support (Table 3.1), and we will go through them from least desirable to most desirable. This progression often corresponds to the expense of the third-party code and libraries, but not always. Also, note that there are other factors involved in making a decision.

The least desirable level of support is none. With this, you end up with the code as is, requiring any future modification to come from you alone. This does not mean that the code is useless, just that if there is a choice with more support that is not substantially more expensive then that is the choice to favor. You must also be careful when integrating the code into your code, particularly if the source is not available to fix problems if they occur.

TABLE 3.1   Levels of Technical Support

Type of Support	Description	Value	Typical Cost
No support	The library is completely your responsibility.	None	Free
Source available	Access to view and, preferably, edit the source code.	High	Varies widely
Community support	A community of people available for questions and some requests.	Moderate	Free
Developer contact	Contact with the developer for answering questions.	High	Moderate
Developer requests	The right to request bug fixes and features from the developer.	Excellent	High

The next step up in support is community-level support, which is a common form of support for open source projects. This type of support is useful primarily if the code and libraries have a large user base. Communication is achieved through a message board or mailing list, and is therefore unpredictable. However, if the community is large enough, there is a better than average chance that another member of the community has already encountered and fixed a problem you are working on. It is a good idea to check out the message board or mailing list before adopting a product with only this level of support in order to see what level of response you can expect. In addition, try to find out if the developers of the code and libraries are regular posters to the message boards or mailing lists. This is a decent level of support when the community is reasonably large.

Another step up in support is direct contact with the developer of the third-party code, which is common for commercial libraries. Optimally this would give you a line of communication to the programmers who are working directly with the code, but this is often not possible. You should at least check to make sure the technical support team consists of programmers and software engineers, since you do not just want someone reading from a script. The best asset to have is access to other programmers who have used the same developer. Question them on the responsiveness and level of support they received. Failing that, see if you can get an evaluation period with the code and come up with some questions pertinent to your project. You can then test the technical support with the questions to see if they meet your expectations.

Finally, we come to the most desirable level of support, and therefore the least common: the ability to request changes to the code or library directly from the developer. As with developer question and answer support, you should determine the level of competence of the developer. With a competent developer, this level of support provides you with extra programmers to solve the subset of problems related to the third-party code. It is necessary to go beyond this, however, and determine what support and additional implementations the developer is legally bound to deliver to you after initial purchase. Remember, we are not talking about requesting a feature for a future implementation, but a direct and timely implementation of a requested feature to meet your deadlines. Be sure both parties understand the exact obligations in order to properly assess the risks involved, and perform necessary adjustments to the schedule.

Research

The best preventative measure to take that will both reduce risk and assuage your personal worries is research. The more information you have available to make the final decision about using an external library, the more comfortable and sure you will be when you do make it. As with almost any activity in software development, there is a danger of becoming mired in continuous research. Therefore, you must strike a balance between the amount of time you spend researching versus the risk of adopting a technology and throwing it away if it does not work. Generally, you will find that this leaves you with plenty of time to perform research.

So, what should you research? Technical support, as we just discussed, is perhaps one of the most important considerations, but certainly not the only one. Sharing the top status with technical support is the applicability of the code or library to your project. Before you can properly research this, you must determine the functionality that you require. Look over your current project plan and group similar technologies together. You might end up with some technologies in multiple groups, which is fine in the research stage. Just remember not to use more than one of them in the final implementation. For each of these groups, make a detailed list of needed functionality, and then go back over it and prioritize the list. Next, determine the functionality most likely to change and put that on a separate list. At this point, you are ready to look at the features of the code or library you are considering adopting. There usually exists a feature list, and even better, sometimes the interface is published. Compare this and mark any missing or vague items on your list. Follow up with an e-mail to the developer asking about these points, which is also a good test of their responsiveness. If the functionality meets your needs or is close enough that you can adapt to it, you should add it to your possible solutions for that problem. Step away from it for a short period before coming to a final decision. In addition, there are a few more considerations to add to that final decision.

The next item to check is whether you can obtain the full source code or only the interface and binary libraries. If you obtain the source code, then even if technical support fails you still have options available to fix the problem yourself. Without the source code, you might still be able to find a workaround, but it will not be as easy. With the source code, it is not only much easier to find a workaround, but you can also change the source when an external workaround is not available. Another advantage occurs if you can obtain at least part of the source code during evaluation. This is another indicator as to the technical experience of the developer. In summary, the lower the support level the more you need the source code, but even with the highest support level there are still benefits to having the source code.

Finally, you will also want to look at the competence of the developer from an overall perspective. The best approach to this is speaking with customers who have used the code or library directly. While it is also useful to read message boards and reviews, nothing beats a direct one-on-one conversation. Do not be afraid to ask the developer for references to others using their product, but try to find some on your own as well as the developer choices might be biased. A large user base is a usually a good sign, and it also makes finding other users to talk with much easier.

GO TO THE SOURCE

When using network code from another party, a project team ran into a problem when one of the library functions was leaking memory, thereby causing the application to fail after only a few calls to that function. After a day of useless investigation, one of the programmers decided to ask if the source code could be obtained. It was provided readily, and within an hour the problem was located. Another hour was all it took to inform the original developer and update the library. The lesson taken from this is to obtain the source code with the library, if it is available. This can save you valuable time if problems occur.

Additionally, if the developer has been around for a considerable amount of time they are likely to be more experienced as well as have a larger user base.

After investigating all the relevant information, you should weight it according to the requirements of your project. At this point, you should also be able to assess the risks involved and the corresponding rewards for taking these risks. Put this all together and make a final decision.

Flavors: Types of Reusable Code

It is important to understand the different types, or flavors, of reusable code that you will encounter. Each type has its own particular uses and corresponding advantages and disadvantages. What follows are the most common types, but do not be surprised if you encounter others on occasion.

Snippets

Code snippets are small pieces of code that are useful but not easily categorized and placed into a library. These tend to come from one of several main sources and are usually free. For example, you might have written a small template that deletes and sets a pointer to NULL in C++:

```
template <typename t_Type>
void delete_and_null(t_Type &io_pointer)
{
    delete io_pointer;
    io_pointer = 0;
}
```

This does not really fit in any library that you have created, so you just bring the small code snippet over to your next project. This might continue through several projects.

Another source of code snippets is fellow programmers. These snippets are generally passed around when one programmer hears another talking about how to solve a problem. The programmer might then offer a small code snippet that he used to solve the same problem.

Finally, there are a couple of public sources for code snippets. Magazine and Web articles will have some useful code snippets. There are also online databases, such as *www.sourcebank.com*, that have searchable repositories of code snippets. Both these sources are worth checking if you are looking for a small piece of code. Be sure to check if any copyrights or licensing applies, but because of the small size, these snippets do not usually have these requirements.

Because code snippets are small single purpose chunks of code, they are generally easy to integrate and remove when necessary. This makes them a very low risk for reuse. The one caveat is that they should still be well documented in the project documentation to avoid duplication of functionality by other team members. Look at the CAP Epidemic to learn more about this danger and how to avoid it.

Standard Libraries

When we talk about using external code, we are not referring to higher-level languages. However, we are referring to the standard libraries that are available for those languages. As with code snippets, these are usually free and place no restrictions on their use. Even better than code snippets, they tend to have a larger user base from the language community for testing, debugging, and feature requests. This makes them robust libraries that are generally useful for practical problems. Because of their availability and large community support, it is also easy to transfer code written using the standard libraries between projects, since the other project can be expected to use the standard library. Another advantage with many of the standard libraries is the tendency for their creators to be experts in the particular language for which the library is designed. Disadvantages tend to come in the support area since there is generally no dedicated support team, but the community support is very good and with all the other advantages makes adopting standard libraries a must for most languages.

Other Libraries

Third-party libraries that are not part of the standard language are where most decisions regarding using external code will have to be made. These range greatly in purpose, quality, cost, and support. This is where you must perform diligent research in order to decide if you want to adopt a library for use with your project. The common format is a set of interfaces to an internal implementation of the libraries' function. Depending on the provider, you might or might not have the source code for the implementation. It is generally better if you can obtain it, but not required.

Properly incorporating the library into your development process is very important. Make sure the documentation for the library is available to the entire team. A good way to do this is have an internal documentation Web site with both the project documentation and documentation for all external libraries and tools. Additionally, the libraries should be added to the build process transparently, or in other words, without making it more complicated. This is easiest if you have one person responsible for integrating libraries into the build process. Once this is accomplished, you can begin to use the library within the project code. Later we will talk about safe ways of using library code within your code base that minimizes the risk of later problems.

Frameworks

Frameworks differ from libraries primarily in where they integrate into program flow. Unlike a library, which your code calls, a framework calls your code. If you do decide to use a framework for your application, chances are that you will only use one framework for the main project. However, some user interface components, such as dialog boxes, might act as miniature frameworks by using callbacks into your code. This means that you can think of the user interface as a framework as well. Many times, the main framework and the user interface framework are the same framework.

Unlike libraries, frameworks are tightly integrated and often hard to remove. They also require early adoption to be of use. Frameworks should be given the most research time, as they can cause the most problems if you change your mind. Because you have less control over the flow of the application with an external framework, it is more important than ever to obtain either source code or the highest level of technical support.

In the case of frameworks, often the user must adapt to the framework more so than adapting the framework for the user's specific purposes. To make this clearer, let us look at one such framework that is a choice available for doing development for cell phone applications. This framework is an essential part of the Java2 Micro Edition (J2ME), and provides both an application and user interface framework. Let us start with a discussion of the application framework. In order to create a

MICROSOFT FOUNDATION CLASSES

One very common framework is Microsoft Foundation Classes (MFC), which has been used in many applications. The MFC framework uses a Document/View structure for the majority of its applications. This breaks the application into separate data and user interface modules. With the help of a project creation tool in Microsoft Visual Studio, the basic classes are created with slots for the user to add application-specific code to each of these modules. The framework handles many common tasks with only minimal implementation required by the programmer using it. A plethora of other supporting classes is provided to aid in the creation of the application.

While MFC is great for certain types of application, particularly small tools requiring user interfaces, it does show some of the common limitations of frameworks. MFC removes the need to implement common tasks, but if small changes are required to these common tasks, it is often difficult to achieve them without considerable work. Sometimes it is even necessary to re-implement certain parts of the framework to achieve these minor modifications.

Another difficulty encountered with frameworks is in integrating them later in development. A project had created an editor tool for editing application-specific configuration files. The tool was originally implemented with the standard Windows libraries, but a suggestion was made to update it using MFC to make new feature implementation easier. Unfortunately, the amount of work that would have been required to update the current tool would have been as much or more than writing a new tool from scratch. This course of action was therefore abandoned.

J2ME application, you must derive at least one class from `MIDlet` and implement three abstract protected methods. This means that your class must have the following signature:

```
public class MIDletImplementation extends MIDlet
{

    protected void startApp()
        throws MIDletStateChangeException;

    protected void pauseApp();

    protected void destroyApp(boolean b)
        throws MIDletStateChangeException;

}
```

Of course, for the application to actually work, the methods must be implemented. Now the environment that this application is run on, be it an emulator or a cell phone, makes calls to the three functions according to a certain set of rules. To initiate the application, startApp is called first. After that, the framework might call pauseApp or destroyApp at any time. If the application is put into the background for some reason, pauseApp will be called. When the application is then subsequently brought into the foreground, startApp will be called again. Notice that the framework leaves it up to the implementer to determine if startApp was called for the first time or to reactivate from a paused state. When the application is terminated from a source external to the application, destroyApp is responsible for handling this. The argument provided to this function gives the implementer the opportunity to prevent the destruction in some cases. Again, the implementer must adapt to both cases.

Now let us look at a sample application, which is somewhat contrived for the purposes of simplicity:

```
public class MIDletImplementation extends MIDlet
    implements CommandListener, Runnable
{

    protected void startApp()
        throws MIDletStateChangeException {
        List list = new List("Menu", List.IMPLICIT);
        list.append("Start", null);
        list.append("About", null);
        list.append("Exit", null);
        list.setCommandListener(this);
        Display.getDisplay(this).setCurrent(list);
    }

    protected void pauseApp() {
        Display.getDisplay(this).setCurrent(null);
    }

    protected void destroyApp(boolean b)
        throws MIDletStateChangeException {
    }

    public void commandAction(Command command,
        Displayable displayable) {
        switch(
```

```
                    ((List)displayable).getSelectedIndex()
            ) {
                case 0:
                    new Thread(this).start();
                    break;
                case 1:
                    Display.getDisplay(
                        this).setCurrent(
                            new Alert("MIDlet Example"),
                            displayable);
                    break;
                case 2:
                    try {
                        destroyApp(true);
                    } catch(
                        MIDletStateChangeException e
                    ) {
                        System.out.println(
                            e.toString());
                    }
                    notifyDestroyed();
                    break;
            }
        }

    public void run() {
        Alert alert = new Alert("Running...");
        // 1000 milliseconds = 1 second
        alert.setTimeout(1000);
        Display.getDisplay(this).setCurrent(
            alert,
            Display.getDisplay(this).getCurrent());
    }

}
```

This is a complete application, but you might have noticed that commandAc-
tion and run are never called. This is because they both represent calls made by the
J2ME framework rather than by the implementer. To understand how these are
used, let us look at the following line first:

```
list.setCommandListener(this);
```

This tells the framework that this object will be handling actions taken by the user on the user interface element `list`. This means that the display and some interactions with the user interface are handled by the framework, leaving the implementer with control only when certain actions are initiated that cause a call to `commandAction`.

The `run` method is eventually called after the following line is executed:

```
new Thread(this).start();
```

Although the implementer has more control over this, the exact timing of the call is controlled by the framework scheduler, or in some cases, the framework might defer to a hardware scheduler. As you can see from this example, the implementer must respond to requests from the framework to handle application-specific functionality. In addition, the user can only exert a certain amount of control over how the framework behaves. Frameworks can be very useful, but it is important to know their limitations even more so than for third-party libraries.

Active Libraries

Although optimization should not be a major concern until it is necessary to optimize, there is certainly no reason not to use a library with better performance if it offers the same features and ease of use that another, slower library offers. Since you are unlikely to modify or maintain the external library yourself, the concerns about Premature Optimization are not as justified when determining which library to use. Because of this fact, active libraries are a new concept that can be useful to understand when choosing a third-party library. However, do not use this line of reasoning to discount third-party libraries altogether, only to choose between different options.

Active libraries are part of the concept of generative programming, which is discussed in more detail in Chapter 2, "CAP Epidemic." Here we are concerned only with the active libraries themselves. An active library is essentially a code generator rather than a fixed code library. This means that an active library generates code specific to the feature requests for an application at compile and link time, rather than determining what code to execute at run time. This provides several optimization advantages, including better performance due to reduced need for conditionals and often less memory due to the inclusion of only code that is necessary for the final application. Assuming the active library provides the same features, they are the preferred choice over fixed code libraries. In fact, because of the reduced code size of the final application-specific library, active libraries can provide more functionality without the associated risk of an overly bloated binary library.

While much of generative programming represents technology that still has a considerable amount of time to mature, active libraries can be implemented in some languages with current techniques. In particular, C++ templates offer a method of creating active libraries that require no extensions beyond the language standard. Templates can be used to execute algorithms at compile time through a variety of techniques that fall under the moniker of template meta-programming. Excluding the limitations of the compiler, template specialization allows the use of conditional compilation and looping code generation through recursion. This forms a Turing complete language just like C++ that in principal can perform any algorithm, except templates can do this at compile time. Due to compiler limitations and the higher complexity of writing templates, this ideal cannot be fully realized. Nevertheless, enough is possible to allow the creation of template libraries that generate code at compile time, making them active libraries as well. Therefore, if you are evaluating libraries, look on template meta-programming as providing an advantage over run-time libraries. If you are writing a library, consider using template meta-programming over conditionals to capture feature variations that only need to be distinguished at compile time and not run time.

To help show the flexibility that active libraries can provide, here is the definition of a debugging flag class that could be part of just such a library:

```
/**    Generic policy-based debug flag template class.
 *     This is a customizable flag intended for
 *     use during debugging. A disabled version that
 *     always returns false is provided for
 *     release builds.  A decent optimizing compiler
 *     will eliminate code that is surround by the
 *     false statement in a release build.
 *
 *     @param    init_policy    - Initialization
 *                        policy class type.
 *     @param    assign_policy    - Assignment policy
 *                        class type.
 *     @param    storage_type    - Storage type.
 *
 *     @par Initialization policy must provide:
 *     @li        typedef <I>type</I> type;<BR>
 *                This is the argument type, so make
 *                it a const reference if it is a
 *                complex type and performance
 *                is a concern.
 *     @li        static storage_type
```

```
*                   convert_to_storage_type(type);
*
*     @par Assignment policy must provide:
*     @li         typedef <I>type</I> type;<BR>
*                 This is the argument type, so make
*                 it a const reference if it is a
*                 complex type and performance
*                 is a concern.
*     @li         static storage_type
*                 convert_to_storage_type(type);
*
*     @par Storage type must support:
*     @li         conversion to boolean
*/
template <
    class init_policy       = boolean_flag_policy,
    class assign_policy     = boolean_flag_policy,
    typename storage_type   = const bool
>
class flag
{
    // ...
};
```

This class is meant to represent a boolean flag that can be initialized in a customizable manner. To understand how this works, let us look at the `init_policy` and `assign_policy`. These two template parameters can take any class that follows the interface guidelines specified in the comments. This allows the initialization function for this class to be written as follows:

```
/** Create new flag.
 *    @param    i_initialState - Value to be
 *                        converted to initial state.
 */
explicit flag(init_type i_initialState)
    : m_state(
        init_policy::convert_to_storage_type(
            i_initialState))
{
}
```

Because the initialization is provided as a template parameter, no object creation or branching is required at run time. This amounts to the same effect as if a code generator had created the required code based on a feature description provided by the user. By predefining the most common policies, the library can offer easy generation of different flags by specifying the features required as template parameters. For example, the simplest policy is to emulate a standard boolean assignment:

```
/// Default boolean policy for use with
/// debug flag.
struct boolean_flag_policy
{
    /// Policy type used by template.
    typedef bool type;

    /** Conversion function from policy type
     *   to storage type. In this case there
     *   is no conversion so the value is just
     *   passed along.
     *   @param    i_value - Value to convert
     *                          to boolean.
     *     @return    Boolean result of conversion.
     */
    static inline bool
        convert_to_storage_type(type i_value)
    {
        return(i_value);
    }
};
```

This is just a small sample of the flexibility that an active library can provide without sacrificing flexibility or performance.

Expanding the Selection with .NET

Programmers tend to favor a particular language for their development purposes. Reasons vary greatly, from the language being the first they learned to a careful evaluation of their needs and programming style. For many, their adherence to a particular language is almost religious in nature, with their own little wars occurring on the message boards and at conferences. Despite this fervor, these programmers have many valid reasons for using one language or another. The real problem is that these languages do not communicate well, and what little communication that does exists must be generated from scratch for each language combination. This means that libraries written in one language are useless to another language, and although not directly caused by NIH Syndrome, it leads to the duplication of effort that is symptomatic of NIH Syndrome.

The introduction of .NET by Microsoft offers a solution to this problem. While it will probably not put an end to the language wars, it does make available a peaceful channel of trade between the warring parties. A library written in any .NET-compliant language can be used by any other .NET-compliant language, with all the work concentrated in making the language .NET compliant. There is no need for custom programming to integrate with each individual language. This, along with a well-formed object model, allows sharing between languages with ease previously unavailable.

The details of .NET require a book in their own right; in fact, many already exist, so we will only take on the task of how this affects NIH Syndrome. The most important impact is the expansion of available third-party libraries that are coming into existence or being updated to meet compliance with the .NET standards. This should provide an unprecedented level of code reuse and third-party library support, but there are still some roadblocks to this new technology. Once again, NIH Syndrome rears its ugly head as many programmers reject .NET technology out of fear.

What are these fears, and how realistic are they? There are two primary fears, both of which are overemphasized by most programmers. The usual suspect is here in the form of performance concerns, and as before, the same rules of avoiding premature optimization apply. In fact, the opposite is true, as .NET can provide a clear development advantage in this area that was previously not available. The application can be written in a safer and easier development language with the corresponding increase in development time, then particular modules that represent performance bottlenecks can be rewritten in a more performance critical but harder to use language to meet performance goals. This is facilitated by .NET because interfacing the two languages is relatively easy.

The other common fear is that .NET reduces all compliant languages to a single common denominator, essentially forcing every language into one. This is a misunderstanding of .NET's purpose that dissuades many from its use. The real purpose of .NET is to provide a common interface between languages that allows the communication of higher-level constructs such as objects to survive. Behind this interface, the language might exist as it always has and take advantage of any constructs unique to that language. Additionally, some constructs, such as multiple inheritance, can be mapped to the .NET language model with only minor concessions. The requirements of .NET are only that the accessible pieces of the language be mapped by the compiler to the .NET model. Thus, languages such as Eiffel remain largely intact even after integration into the .NET environment.

As with any technology that interfaces between different models, there is overhead to this connection. However, as with other performance concerns, this is unlikely to be problematic in more than a few instances. These instances can then be handled at the end of the project in the optimization phase. This will usually involve the shifting of some responsibility between two different languages and should not present a real problem.

The final and most valid concern about adopting multiple languages lies in support. This can be handled in several ways. A team of individuals working on closely related aspects of a project should stick to one language, as their code is likely to intersect often. If the entire team is just a few people, the entire project code base should stick to one language. However, on large teams that are broken into smaller self-contained teams, it is possible for the teams to develop in different languages as long as the interface and responsibilities are clearly defined. No matter what the size of the team, third-party libraries from any language should be considered. However, the one caveat is the need to more carefully consider support for these libraries. In this case, even if your team can obtain the source code, it must have members who understand the source language; otherwise, it is necessary to look for an increased level of support in case of problems.

Overall, the introduction of .NET should expand the range of libraries available across the software community. However, there is another concern shared by many in the software community, particularly open source developers. Because the .NET technology is proprietary to Microsoft, there is the fear of the level of control that will be available in the future. Before we analyze this fear, it is important to understand some issues about .NET, and how these relate to NIH Syndrome. For Microsoft, the .NET initiative represents a large number of changes including corporate branding. We are, however, really only concerned with the portion that facilitates sharing of object libraries between different programming languages.

This communication between programming languages, and the ability of compiler writers to adapt to .NET, is made possible by a set of open standards called the *Common Language Infrastructure* (CLI). For a language to interact with .NET, it must support a mapping to the *Common Language Specification* (CLS) and .NET object model. This is achieved by an *intermediate language* (IL), which can then be run on an interpreter or *virtual machine* (VM). The IL might alternatively be compiled either ahead of time or with a JIT (Just In Time) compiler.

We can now return to our concern about the proprietary nature of this technology. Because a valid .NET language must compile to the IL, which follows an open specification, then it is possible to create a VM and JIT compiler that can use the IL. Just such a project is currently in progress under the moniker of Mono sponsored by the open source developer Ximian. Mono will be open source and therefore free of any charges or proprietary information. In addition, due to the open nature of the IL, any language can be modified to produce IL code and therefore integrate with either .NET or Mono.

The end advice is to conquer your fears about language interoperability and begin to seriously consider if your project could benefit from the .NET framework. As with every new technology, you must analyze the risks and rewards, but you can expect to see future work to continue in the direction of language interoperability.

Strategy Pattern Revisited

When we looked at preventing premature optimization, we discussed the Strategy Design Pattern [GoF95]. To recap, the basic idea of the Strategy Pattern is to abstract the interface to an algorithm so that multiple algorithms for the same functionality can be encapsulated in interchangeable objects. This pattern is described in detail in the *Design Pattern* book by the Gang of Four, so here we will just examine how it applies to reducing the risks of adopting third-party code and libraries.

When adopting any library algorithm, it is a good idea to encapsulate the algorithm in an object that follows the Strategy Design Pattern. By doing this, changing to a different library's implementation or your own implementation of the algorithm is an easy transition. The important part of this approach is to carefully consider the public interface of the Strategy so as not to expose any library-specific details that will make transitioning difficult.

To accomplish this, it might be necessary to convert objects and values back and forth from your project's representation. This is a good idea in any case, as a common representation for objects across the project is important to communication and sharing of code and responsibilities. For example, the only time you should expect to see more than one representation of a three-dimensional vector is when interacting with an external library. By using the strategy pattern, and the other patterns we will talk about shortly, you can minimize the areas of the code where this occurs. As always, the conversion process will lead to concerns about performance. For reassurance about why this is not a concern, see Chapter 1. In short, when the time comes to optimize, you will be able to sacrifice this conversion if and only if it is necessary for performance, because the code base will be stable and therefore changes will not cause development hassles.

For common conversion operations, it is vital to provide an automated method for converting and handling the different object types of the library and project code. This will encourage the consistent use of conversion by removing the tedium of continually implementing the same conversion. In addition, the presence of the library's object type can be reduced to only direct interactions of the library. To better understand what this means, let us look at an example of two different vector types that might be encountered on a project using another developer's library. The full C++ code for this example is included on the companion CD-ROM in Source/Examples/Chapter3/convert.h and Source/Examples/Chapter3/convert.cpp. Here is the vector class that the project is using:

```cpp
class t_Vector
{
public:

    t_Vector()
        : m_x(0.0f), m_y(0.0f), m_z(0.0f)
    { }

    t_Vector(float i_x, float i_y, float i_z)
        : m_x(i_x), m_y(i_y), m_z(i_z)
    { }

    float m_X() const { return(m_x); }

    float m_Y() const { return(m_y); }

    float m_Z() const { return(m_z); }

private:

    float m_x, m_y, m_z;

};
```

Here is the vector class used by a third-party library:

```cpp
class t_NIHVector
{
public:
```

```
        t_NIHVector()
            : m_x(0.0), m_y(0.0), m_z(0.0), m_w(0.0)
        { }

        t_NIHVector(double i_x, double i_y,
        double i_z, double i_w)
         : m_x(i_x), m_y(i_y), m_z(i_z), m_w(i_w)
        { }

        double m_X() const { return(m_x); }

        double m_Y() const { return(m_y); }

        double m_Z() const { return(m_z); }

        double m_W() const { return(m_w); }

private:

        double m_x, m_y, m_z, m_w;

};
```

Notice that there are two major differences between the two vector classes. The project class uses float, whereas the library uses double, and the library vector contains the extra w coordinate. To automate the conversion process, we need an object that can be assigned to from both types and then used as both types:

```
class t_NIHVectorProxy
{
public:

    explicit t_NIHVectorProxy(
        const t_Vector &i_vector)
        : m_vector(
            i_vector.m_X(),
            i_vector.m_Y(),
            i_vector.m_Z(), 1.0)
    { }

    explicit t_NIHVectorProxy(
        const t_NIHVector &i_vector)
        : m_vector(i_vector)
```

```
    { }

    t_Vector m_Vector() const
    {
        return(t_Vector(
            static_cast<float>(m_vector.m_X()),
            static_cast<float>(m_vector.m_Y()),
            static_cast<float>(m_vector.m_Z())
            ));
    }

    t_NIHVector m_NIHVector() const
    {
        return(m_vector);
    }

    const t_NIHVector &m_ConstNIHVector() const
    {
        return(m_vector);
    }

private:

    t_NIHVector m_vector;

};
```

Notice that we store the vector as the more complex of the two types. This is to prevent loss of the extra information contained in the w coordinate. Even with this precaution, it is still easy to accidentally lose this information when performing conversions. An alternative method, known as the Adapter Pattern, exists that can further reduce this possibility, but even it cannot eliminate this danger. Therefore, it is more important than ever to comment conversion and adapter classes. We will look more at the Adapter Pattern later; for now, let us look at how we could use our conversion helper. First, let us imagine that we have two library functions for getting and setting a location as follows:

```
t_NIHVector m_GetLocation();

void m_SetLocation(const t_NIHVector &i_vector);
```

In addition, we need a function in our project with which we can add two vectors:

```
t_Vector m_VectorAddition(
    const t_Vector &i_lhs,
    const t_Vector &i_rhs)
{
    return(t_Vector(
        i_lhs.m_X() + i_rhs.m_X(),
        i_lhs.m_Y() + i_rhs.m_Y(),
        i_lhs.m_Z() + i_rhs.m_Z()
        ));
}
```

Now we can write a function that moves the library location by a delta vector:

```
t_Vector m_Move(const t_Vector &i_delta)
{
    t_NIHVectorProxy
        l_location(m_GetLocation());
    t_NIHVectorProxy l_newLocation(
        m_VectorAddition(
            l_location.m_Vector(),
            i_delta));
    m_SetLocation(
        l_newLocation.m_NIHVector());
    return(l_newLocation.m_Vector());
}
```

Notice that the m_VectorAddition function returns a new object of type vector, which therefore cannot have retained the w coordinate of the library location vector. In this instance, we know that the w coordinate is always 1.0, but if we could not guarantee this, we would have to find another method for adding the vectors or saving the w coordinate. This illustrates the dangers of mixing types and why they must be isolated as much as possible. To this end, let us look at some other useful techniques for containing the library interactions.

Adapters, Bridges, and Façades

While the Strategy Design Pattern works well for isolating individual third-party algorithms, you might be adopting a more full-featured set of code and libraries that represents more than just an algorithm. Fortunately, other useful design patterns can be used to isolate these larger concerns from the project code to assist in case of the necessity for changing libraries. As with the Strategy Pattern, the *Adapter*, *Bridge*, and *Façade* Patterns can be investigated in detail in the Gang of Four's *De-*

sign Patterns book. Here we will discuss the specifics of their use in providing a layer of protection from external code.

Bridge Pattern

We start with the Bridge Pattern, which is meant to separate an implementation from the corresponding abstraction that it represents. This applies perfectly to the use of external libraries, where you have an existing implementation with the possibility of changing to a different implementation if problems arise.

To create a bridge, you must examine the library and determine what functionality your project is planning to use. With this in mind, construct an interface that conforms to the style of your project. The key concept to apply while constructing this interface is making it minimal but complete. Do not add a lot of helper functions or access to unrelated parts of the library. These can be added later at a different layer or through a different bridge, and will only confuse the use and maintenance of the abstraction you are initially creating. On the opposite side of the coin, you should try to anticipate any reasonably minor changes and likely functionality that is related to the parts of the library being used. Creating the interface is a balancing act between these two opposites that is unique for each new project and library.

With the interface designed, the bridge can be connected to the library and placed into the project for use. Figure 3.1 shows the relationship of the library to the project using the Bridge Pattern. The library implementation should be private and remain inaccessible except from the bridge or bridges for that library. With most languages, this must be accomplished with project conventions since it is difficult to restrict access to a library meant for reuse.

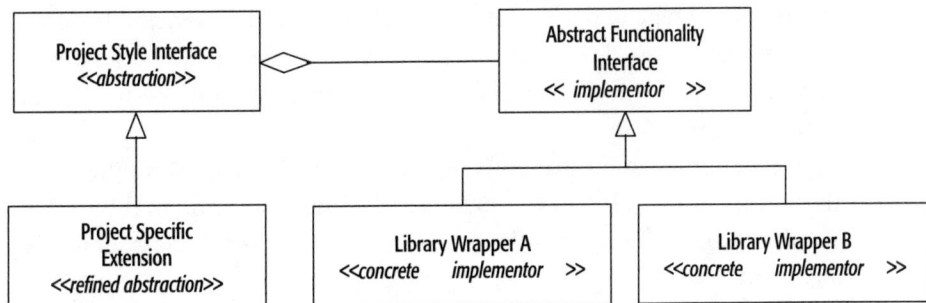

FIGURE 3.1 A bridge between the project style interface and the libraries is made through a private interface that provides uniform access to each library's wrapper.

An additional advantage is that the implementation can vary even at run time, allowing experimentation with different libraries for evaluation purposes. However, as with all abstractions there is a performance penalty implicit in the redirection necessary to create the bridge. This should only be a concern at the end of development in the optimization phase, and at this time, the bridge can be removed without the problematic consequences of not creating it early in the project lifecycle or removing it too early.

Adapter Pattern

In any object-oriented language, there is a good chance that a library will return an object with information you need to save, access, and pass back to the library. If your project has not already planned an object of this type, you might consider adopting the library's implementation. However, if you already have an object of a related type, it is not wise to have two related but differing representations in the project code base. A common example is a vector or matrix representation, which has been redeveloped more times than many of us would like to hear. You therefore must come up with a way for your project to interact with its type and the library to interact with its type seamlessly.

One method for accomplishing this, which was mentioned earlier, is to convert objects from the library to objects in your project. However, this approach is not sufficient in all cases. Suppose, for instance, that the library representation contains an extra bit of information that your type is not capable of representing. To solve this, you could create a container that holds your type, converted from the library type, and the extra information, but this becomes cumbersome to manipulate and track.

A better solution is to use the Adapter Pattern to wrap the library object in an interface that is shared with your object type. You can then present an object that acts like your type while maintaining any internal data specific to the library implementation. This also facilitates changing library implementations as you already have the adapter interface available and only need to wrap the new libraries object in the same interface.

To better show the tradeoffs involved, let us look at an alternative solution to the vector conversion that was mentioned in our discussion of the Strategy Pattern. For this example, we extend the functionality of the vector class slightly to allow assignment:

```
class t_Vector
{
```

```
public:

    t_Vector()
        : m_x(0.0f), m_y(0.0f), m_z(0.0f)
    { }

    t_Vector(float i_x, float i_y, float i_z)
        : m_x(i_x), m_y(i_y), m_z(i_z)
    { }

    float m_X(float i_x) { return(m_x = i_x); }

    float m_X() const { return(m_x); }

    float m_Y(float i_y) { return(m_y = i_y); }

    float m_Y() const { return(m_y); }

    float m_Z(float i_z) { return(m_z = i_z); }

    float m_Z() const { return(m_z); }

private:

    float m_x, m_y, m_z;

};
```

As well as the library vector class:

```
class t_NIHVector
{
public:

    t_NIHVector()
        : m_x(0.0), m_y(0.0), m_z(0.0), m_w(0.0)
    { }

    t_NIHVector(double i_x, double i_y,
        double i_z, double i_w)
        : m_x(i_x), m_y(i_y), m_z(i_z), m_w(i_w)
    { }

    double m_X(double i_x) { return(m_x = i_x); }
```

```
    double m_X() const { return(m_x); }

    double m_Y(double i_y) { return(m_y = i_y); }

    double m_Y() const { return(m_y); }

    double m_Z(double i_z) { return(m_z = i_z); }

    double m_Z() const { return(m_z); }

    double m_W(double i_w) { return(m_w = i_w); }

    double m_W() const { return(m_w); }

private:

    double m_x, m_y, m_z, m_w;

};
```

Now to implement the Adapter Pattern, we must extract an interface from the project vector class:

```
class i_Vector
{
public:

    virtual float m_X(float i_x) = 0;

    virtual float m_X() const = 0;

    virtual float m_Y(float i_y) = 0;

    virtual float m_Y() const = 0;

    virtual float m_Z(float i_z) = 0;

    virtual float m_Z() const = 0;

};
```

This interface will be used when possible in the project code instead of the actual vector class. For example, we now implement the vector addition as follows:

```
void m_VectorAddition(
    i_Vector &io_lhs,
    const i_Vector &i_rhs)
{
    io_lhs.m_X(io_lhs.m_X() + i_rhs.m_X());
    io_lhs.m_Y(io_lhs.m_Y() + i_rhs.m_Y());
    io_lhs.m_Z(io_lhs.m_Z() + i_rhs.m_Z());
}
```

Notice the necessary difference is that we no longer want to create a new vector if possible. This is because the new vector could be a different type than the vector we are trying to update. There are ways to handle even this by using the Abstract Factory Pattern, but for now, let us just use the method presented here. The good news is that larger objects, which are good candidates for the Adapter Pattern, generally do not get instantiated as often, so they avoid this problem.

As you can see, things can become somewhat complicated when restricted to using interfaces. Using the Adapter Pattern can have wider reaching consequences if you are not already using interfaces for the project classes involved. This is the main disadvantage of using the Adapter Pattern, and the only tradeoff that will have a major impact on your decision to use conversion or adaptation.

Moving on, now we can derive the concrete vector class from this:

```
class t_Vector : public i_Vector
{

    // ...

};
```

The most important step comes next, as we make a wrapper for the library vector that implements the project vector interface:

```
class t_NIHVectorWrapper : public i_Vector
{
public:

    explicit t_NIHVectorWrapper(
        const t_NIHVector &i_vector)
        : m_vector(i_vector)
    { }

    t_NIHVector m_NIHVector() const
```

```
    {
        return(m_vector);
    }

float m_X(float i_x)
{
    return(
        static_cast<float>(
        m_vector.m_X(
        static_cast<double>(i_x))));
}

float m_X() const
{
    return(
        static_cast<float>(
        m_vector.m_X()));
}

float m_Y(float i_y)
{
    return(
        static_cast<float>(
        m_vector.m_Y(
        static_cast<double>(i_y))));
}

float m_Y() const
{
    return(
        static_cast<float>(
        m_vector.m_Y()));
}

float m_Z(float i_z)
{
    return(
        static_cast<float>(
        m_vector.m_Z(
        static_cast<double>(i_z))));
}

float m_Z() const
{
```

```
        return(
            static_cast<float>(
            m_vector.m_Z()));
    }

private:

    t_NIHVector m_vector;

};
```

This wrapper contains the library vector, thus preserving any extra information. Finally, we rewrite the m_Move function to use our new functionality:

```
t_Vector m_Move(const i_Vector &i_delta)
{
    t_NIHVectorWrapper l_location(m_GetLocation());
    m_VectorAddition(l_location, i_delta);
    m_SetLocation(l_location.m_NIHVector());
    return(
        t_Vector(
            l_location.m_X(),
            l_location.m_Y(),
            l_location.m_Z()));
}
```

At this point, the Bridge and Adapter Patterns probably sound nearly identical. In many ways, they represent a similar approach to the problem of hiding the library implementation, but there is an important difference. In the case of the Bridge Pattern, the interface you are developing for abstraction comes directly from your use of the library implementation. The Adapter Pattern, on the other hand, is meant to adapt a library object to an existing interface already designed for your project.

Façade Pattern

So far, we have talked about abstracting specific functionality for varying the implementation from the abstraction. We can, and should, take this one step further by using the Façade Pattern. This pattern provides a unified interface to a collection of subsystems that make up a coherent module. For example, the rendering system of an application could be composed of several different subsystems with differing interfaces. By placing a façade upon the rendering subsystems, the rest of the application deals with a single interface.

The true advantage of this for external code and library use comes when a portion of the library in a system is lacking in support for a necessary subsystem. With the façade in place, it is easier to add a separate library into the mix without providing access to the library that would result in confusion about which library is responsible for what functionality. This reduces the risk of confusion by ensuring that only a few people or a single person is exposed to the problems of combining multiple implementations.

The façade should be a second layer of abstraction for external libraries, with the Bridge Pattern forming the first layer of abstraction. This provides the maximum flexibility for adapting to changing requirements and unexpected problems with the third-party code. This also allows you to get the most out of external libraries without the fear of everything depending on them. Even if one portion of the library fails, it is isolated and can be replaced individually with no effect on the rest of the library or the project code.

All these patterns and other preventative techniques aim to reduce the risk involved in using code and libraries from other programmers and developers. With reduced risk comes reduced fear, and this removes the main force behind NIH Syndrome. Even if you do not possess the usual fear of using others' code, these techniques can benefit you and protect from the inevitable unexpected problems that arise in software development.

CURE

What if you have already written a lot of code and ignored what was available from other team members and external developers? Should you continue along this path, leaving your code as is? Let us look at why you might want to consider replacing existing code with external code, and what approach to take when doing so.

Sacred Code

Certain objects and various animals are revered in the many of the world's religions and given special privilege and protection. These objects are shown great respect by the people of that religion, and are sometimes even protected by the laws of the lands in which the religion is practiced. However, the sacred code of programmers, or pieces of personal code that we protect with a religious fervor, does not deserve the same level of respect. Understandably, when we spend a considerable amount of time developing a piece of code we develop a certain attachment to it. This clouds our judgment, and before we can consider refactoring code to use external libraries, we must overcome this prejudice. The best approach to obtaining a more objective look at whether your code has the merit you first ascribe to it is to seek out a third party, preferably someone not directly involved with the section of code in question. Have this person play the devil's advocate and offer you an opportunity to defend your code. This allows you to come out reassured that your code is up to the challenge, or in some cases, you will come out with the realization that your code is not appropriate and should be replaced.

However, there lies a danger in ignoring the sacred code of programmers. Just as with many sacred objects of religion, many assume that the taboos associated with the object are illogical. Many see the believers in the religion suffering, yet refusing to break a taboo, and wonder what kind of insanity has possessed them. However, there is a meaningful explanation for this behavior. Many of these objects represent a source of practical usefulness. Some objects have uses to the people who worship them that are not immediately obvious. This is all overlooked by many outsiders because of the different culture and environment from which they come [Harris74]. Breaking the taboos of these religions would be disastrous, as would ignoring the sacred code of programmers without first understanding their purpose from the point of view of the developers who created them. The lesson to take away from all of this is that you should be wary of sacred code kept around for no valid reason, but be sure to query the programmer who is protecting the sacred code carefully to ensure that it is not just a case where you do not understand the reason.

When to Refactor

Once we are over our reticence, when should we consider replacing our own code with an external library or code snippet? To answer this, first consider when *not* to replace existing code. If the code is fully functional and thoroughly tested, you will not want to replace it with other code. Outside of this, there is room to consider third-party libraries on a case-by-case basis. Unfortunately, there are no set rules for determining when to perform this replacement. The best approach is to estimate the amount of time it will take you to complete the required extra functionality versus the time to integrate the library and replace existing functionality.

One advantage of refactoring that is often overlooked in this case is the long-term benefit of performing the refactoring. It is common to only consider the short-term benefits, but this can skew the results in favor of not refactoring. There-fore, it is a very good idea to look at the future functionality that is on the schedule and any functionality that has a good chance of being added or modified. By doing this, you might find that the library provides a better alternative to the current code for performing this future functionality. By performing the refactoring early, you will save development time as each of these tasks arises.

One Step at a Time

Once you have decided to switch out your code for new third-party code, you can begin refactoring. This refactoring must be performed at a slow and considered pace. Each basic unit of functionality that currently exists in your project should be replaced one at a time. After each swap, you must perform your automated tests to ensure that the functionality has not been changed by the new library calls. You should already have these tests in place; if not, now is a very good time to generate them. They must be in place before the code is changed. This makes it much easier to test for functionality changes because you can perform regression testing, which is done by comparing the results of tests performed before changing the code to those performed after. If the tests match, the chance that problems have been in-troduced by the change of libraries is greatly reduced.

If a test does fail, then your best approach is to locate the failed test and isolate the cause. If possible, this will allow you to reduce the code that exhibits the error to a smaller set of code. By doing this, you not only understand the problem bet-ter, you have a transferable example you can use to obtain help from your avenues of support. Be sure to verify that the library is at fault and not your own code. Have another programmer look at it if possible to gain objectivity. This also has several advantages, chief among them that you can provide the support resources with information on what you have already determined about the problem and what is not contributing to it. A second advantage is the good relations that you can build with support by not making them chase problems that you could have solved.

Conformance

Another important refactoring exercise to perform exists when you are using ex-ternal code directly, rather than through a library. Because there is a wide variety of coding standards and approaches to problem solving, it is unlikely that code out-side of your project will follow the same guidelines that you do. When you integrate libraries, this difference is isolated by the Bridge Pattern. However, when integrating code directly, such as with snippets of code, it is unlikely that creating a wrapper is necessary or even useful.

Instead, you should refactor the code to meet your coding standards. It is a good idea to integrate the code first before refactoring and ensure that it works, also giving you the opportunity to build the necessary test cases. Once integrated, you can begin refactoring the code one step at a time. Continuous testing ensures that functionality does not change. The new code will now be easier for other team members to understand and modify. Without this refactoring, each team member that had to interact with the code would have to work through the code style in order to determine what it was doing. This will cost development time, particularly if the code requires many changes across the course of the project. A side benefit of this process is a better understanding of the code that was just added to your project.

Workarounds

Another important consideration to take into account when using third-party libraries is how to handle problems and missing features in the libraries you are using. The first step to solving these problems is not to throw out the library. Instead, you should follow a set of steps that can allow you to continue to use the library and not lose all the work done so far.

You should start by verifying that you are using the library correctly. Carefully read the documentation related to the problem, and ensure that all reasonable variations have been tried. If you are then satisfied that the problem lies with the library, you should create a simplified version of the problem. This serves two purposes. The first purpose is to clarify the problem in your own mind so you can be sure nothing was missed. The second purpose is to provide a simple example that can be sent out for external support. Once you have constructed the example and confirmed that it suffers the same problem, send it out to all the technical support sources you have for the library. This might include the developers, message boards, and mailing lists. Be sure to explain what you want to accomplish, and ask if there is a current solution or, if not, a workaround.

While waiting for a response, you can consider your own workaround possibilities. Once you receive the response, take the proposed solutions from the responses and your own workaround solution and determine which is the best. If the solution ends up being a workaround, and not just something that you missed in the library, you need to take several more actions to reduce the future risks. First, implement the workaround with clear documentation. Preferably, a central list of workarounds should be available and read by all team members. This will allow you to easily locate and resolve any changes related to the workaround. Next, if possible, correspond with the developers and request that the problem be fixed in a future release. Even better, if you have full support you can request that an immediate fix be implemented. When the fix is available, you can remove the workaround to improve the robustness of the code.

Did We Mention the Importance of Testing?

If you have read the first two major illnesses, you understand how testing is important to preventing and curing those illnesses. Once again, testing is an essential part of handling NIH Syndrome properly. We already mentioned one of the important uses of testing when we discussed the best approach to performing any NIH Syndrome refactoring. There are two other contributions testing can make to reduce the risk of adopting third-party software.

First, tests allow for confidence in the event of switching libraries or changing to an internal implementation. A solid set of tests should be in place from the beginning of library integration. These tests should reflect the needs of the project, not the full capabilities of the library. If you go beyond the needs of the project, other libraries might fail the tests even if they are suitable for the tasks your project requires. With these tests in place, you can easily switch libraries and then run the tests to assure that the functionality remains identical. Without automated tests, the application must be tested by hand after such a large change. Manual testing is tedious and difficult, as well as prone to oversights.

Second, a form of development known as test-driven development is the basis for another technique that can assist in the evaluation of libraries' suitability for your project. The basic idea is to write the test cases for the functionality required by the project. Ensure that all the important issues are tested. At this time, the tests will not even compile because the functionality is not there. You should also not look at any libraries or code that might be considered for providing the associated functionality. Next, write just enough for the tests to run and fail. This is important because you must test the tests for their validity. Once you are assured that the appropriate tests are in place, you can bring in the library you want to use for the desired functionality. As you integrate the library, you can check to see if each test passes and therefore confirm that the library is providing the expected functionality.

RELATED ILLNESSES

Premature Optimization is an all too common reason behind NIH Syndrome. The fear of obtaining poor performance from third-party code leads many programmers to seek a safe haven in writing their own code. This is not only a fallacy, but also one of the causes of the CAP Epidemic. Duplicate code proliferates the industry, much of it available to the programmers who are duplicating it on their own. The solution to this is to put the concerns about optimization in their correct place at the end of development, and start considering the use of third-party libraries at the beginning of development to avoid duplicating work already done.

NIH Syndrome is one of the many problems born of the myopic planning common to software developers, which is in its own right the minor illness of Myopia. This is not only because writing all your own code ends up being a time-consuming process, but also because features available in external libraries are often perceived as unnecessary until it is too late to adopt them to the project. Do not ignore the certainty of additional features and changing requirements when planning. Otherwise, the project might be built on poor code written by overworked programmers that will eventually collapse like Brittle Bones.

FIRST AID KIT

The most important tools to overcoming NIH Syndrome are the third-party libraries. By using these, you will overcome your fear of external code and learn the proper techniques for evaluating third-party libraries objectively. If you are concerned about quality, it is best to start with heavily used libraries such as the Standard Template Library and Boost for C++. From there, you can work your way toward more specialized libraries suited for your particular application.

Adopting .NET technology is another tool that can be used to increase the number of libraries available for use. You can work your way slowly toward an adoption of .NET, starting by using a language or languages that support .NET, such as C#, Managed C++, or Eiffel, among others. From there it is relatively easy to begin using libraries from other languages and only slightly more difficult to adapt your code to be used by other languages. For platforms outside of Microsoft's domain, look for future technology similar to .NET such as the open source version of .NET known as Mono.

SUMMARY

The symptoms of NIH Syndrome:

- A desire to control every aspect of the application code because of a lack of trust in other programmers.
- Not using external code and libraries because of an unnecessary fear of:
  - Poor performance
  - Unknown learning curve
  - Minor missing functionality
  - Debugging difficulties
  - Affect on deadlines
- Duplicating code and work internally that has already been accomplished externally.
- Heavy reliance on internal development because of the belief that all facets of software engineering can be accomplished better internally.

To prevent NIH Syndrome:

- Use middleware and open source to reduce development time.
- Balance the time spent evaluating external solutions with the savings they provide.
- Be sure the appropriate level of technical support is available as a reassurance that the external solution will not cause undue problems.
- Understand the benefits and disadvantages of code snippets, standard libraries, run-time libraries, active libraries, and frameworks.
- Use .NET to expand the available external solutions.
- Use the Strategy, Adapter, Bridge, and Façade design patterns to provide encapsulation and abstraction of external libraries to allow easier integration and smoother replacement when necessary.

To cure the effects of NIH Syndrome:

- Do not allow internal code to become a sacred cow, protected for no valid reason.
- Do not be afraid to remove internal code in favor of better external code.

- Perform any transition to external code one step at a time using standard refactoring techniques.
- When possible, change or wrap the external interface to conform to internal standards.
- Apply workarounds carefully, and remove them as soon as they are no longer necessary.
- Use testing liberally to verify external code works as expecting and perform regression testing when changing implementations.

PART

II

Minor Illnesses

While not quite as prevalent as the major illnesses, each of these minor illnesses represents a common mistake that inevitably costs development time. Because they occur less often, it is more important to stay vigilant when watching for their symptoms. With an understanding of their nature, you can learn to not only detect them, but also to prevent and cure the problems they cause. Learning to prevent and cure these illnesses can provide valuable refinements to the development process.

4 Complexification

DESCRIPTION

Programmers, who are a thinking breed, require at least occasional challenges to truly enjoy work. This desire to conquer a challenge is essential to the often problem-heavy world of software development, but can become a liability when no challenge is present. Complexification is a common minor programmer illness that appears when only a simple solution is required. The afflicted programmer will ignore or discount this solution, and instead endeavor to make the problem more complex in order to necessitate a more complex solution. The detrimental effect of this line of thinking should be immediately obvious. More work and more risk are brought on by the programmer's need for complexity.

SYMPTOMS

Software development is a modern industry filled with complex problems that require complex solutions. Because complex problems and their corresponding complex solutions do exist, when do we consider a solution too complex for the problem? Let us look at some of the outward signs that might indicate an overly complex solution.

Poor Readability

One of the results of overly complex code is code that is difficult to read. This is only an initial indicator and the reason for the poor readability must be investigated more fully, because poor readability can also be caused by other illnesses or simply poor coding. On the other side of the coin, some readable code is still too complex, but this is less common. If you write overly complex code, you can expect that a majority of the time it will be less readable.

For a good example of how increased complexity decreases the readability of the code, we return to two algorithms first presented in Chapter 1, "Premature Optimization." They can also be found on the companion CD-ROM in Source/Examples/Chapter1/readability.cpp. First, the simple selection sort algorithm:

```
void selection_sort(int *io_array,
                         unsigned int i_size)
{
    for(unsigned int l_indexToSwap = 0;
        l_indexToSwap < i_size; ++l_indexToSwap) {

        unsigned int l_indexOfMinimumValue =
            l_indexToSwap;

        for(unsigned int l_indexToTest =
            l_indexToSwap + 1;
            l_indexToTest < i_size;
            ++l_indexToTest) {

            if(io_array[l_indexToTest] <
               io_array[l_indexOfMinimumValue]) {
                l_indexOfMinimumValue =
                    l_indexToTest;
            }
        }

        int l_minimumValue =
            io_array[l_indexOfMinimumValue];
        io_array[l_indexOfMinimumValue] =
            io_array[l_indexToSwap];
        io_array[l_indexToSwap] = l_minimumValue;
    }
}
```

Now compare this to the heap sort algorithm, which is faster but also more complex:

```
void sift_down(int *io_array, unsigned int i_size,
               int i_value, unsigned int i_index1,
               unsigned int i_index2)
{
```

```
    while(i_index2 <= i_size - 1) {
        if((i_index2 < i_size - 1) &&
           (io_array[i_index2] <
            io_array[i_index2 + 1])) {
            ++i_index2;
        }
        if(i_value < io_array[i_index2]) {
            io_array[i_index1] = io_array[i_index2];
            i_index1 = i_index2;
            i_index2 *= 2;
        } else {
            break;
        }
    }

    io_array[i_index1] = i_value;
}

void heap_sort(int *io_array, unsigned int i_size)
{
    if(i_size < 2) {
        return;
    }

    for(unsigned int l_hire = i_size / 2;
        l_hire > 0; --l_hire) {
        sift_down(io_array, i_size,
                  io_array[l_hire - 1],
                  l_hire - 1,
                  (2 * l_hire) - 1);
    }

    for(unsigned int l_retire = i_size - 1;
        l_retire > 1; --l_retire) {
        int l_value = io_array[l_retire];
        io_array[l_retire] = io_array[0];
        sift_down(io_array, l_retire,
                  l_value, 0, 1);
    }

    int l_swap = io_array[1];
    io_array[1] = io_array[0];
    io_array[0] = l_swap;
}
```

Not only is the second algorithm longer, but the flow is more difficult to follow. Even though the heap sort algorithm is only a step or two above selection sort, it is already harder to read. Algorithms that are even more complex become even less readable.

To decide if code that is difficult to read is a result of Complexification, the purpose of the code must first be determined. This illustrates an important reason for both writing minimal but complete code, and for properly documenting code that must be more complex. Without these indicators, programmers unfamiliar with the code will be forced to manually decipher the purpose of the code. This is particularly time wasting if the code requires no changes.

Part of the process of the understanding the code's purpose is to evaluate the contexts in which the code is used. This knowledge will often show if the complexity of the code is caused by imposing unnecessary requirements. This can be a primary indicator of overly complex code, and once the realization is made that these requirements can be relaxed, it becomes obvious that the code can also be simplified.

Suppose you discover that the heap sort algorithm is being used for a collection of items that only need to be sorted when the application is first started. The primary motivation for the extra complexity in this case is the increase in performance, but it is unlikely that the increased speed is necessary in this case. Therefore, the algorithm used is more complex than necessary.

Once a complete understanding of the code and its requirements is reached, the other symptoms can be investigated to see if the requirements are overly strict. While the following symptoms are common, there are other possible reasons for the requirements to be overly strict and the code overly complex. For example, Premature Optimization represents a major cause of writing overly complex code. Symptoms will often overlap as well, such as optimizations done because of Premature Optimization having so little performance impact that the result is invisible to the end user. If the complexity is caused by one of the more obvious symptoms, the diagnosis of Complexification is easy. However, if it at first appears to fall outside the symptoms it might be a good idea to take a closer look and make sure there are valid reasons for the code to be complex.

Invisible Functionality

The end goal of any software application is to provide the users with some specific functionality, whether it is balancing a checkbook or entertaining them with a challenging game. In order to sell more products, the user must be satisfied with the functionality provided by the application. This satisfaction comes from the perceived experience of using the product and the completion of any goals the user wants to accomplish.

The use of the word *perceived* is important, because it indicates the importance of the result over the internal implementation used to achieve that result. This means that any internal functionality that is not perceived by the user in any substantial way is a poor use of development resources. The most common aspects of an application that can be overlooked by users are minor performance enhancements, and certain elements of the visual presentation. In particular, the efforts wasted and problems caused by imperceptible performance improvements are substantial enough to merit their own major illness. See Chapter 1 for a more in-depth discussion of this problem.

INVISIBLE ASTEROIDS

Imperceptible elements of visual presentation are most often seen in games and other visual simulations where accuracy is not the end goal. It is relatively easy to add computations for visual elements that the end user cannot perceive. To provide a better understanding of what this means, let us look at a project in which a large number of asteroids were being simulated and displayed. When the time came for optimization at the end of the project, it was discovered that a large amount of time was being spent with a function for handling the physical simulation of each asteroid chunk. As it turned out, each asteroid chunk was being individually simulated to take into account such details as momentum and energy conservation. However, after looking at the asteroid belt during execution it was obvious that the movement of the individual asteroid chunks was barely noticeable. Both the work and loss of performance could have been avoided by using a simpler solution such as a series of random animations. At this point in the project, it was too risky to implement a completely new solution. Not all was lost, however, as it was possible to remove some of the processing required by this complex solution. Because the asteroids chunks had no effect other than visual presentation, it was possible to only simulate their movement when they were being displayed. While this did recover some of the performance cost, there was no way to recover the lost development time.

Return of the Nifty Algorithm

When we talked about Premature Optimization, we mentioned that one of the causes was the unnecessary use of algorithms that were just recently learned. There is always a temptation to use a recently learned algorithm, as it is fresh in our minds. This is particularly true when we read about the algorithm as written by the algorithm's inventor, who will espouse the superior properties of their algorithm. Although we first mentioned it when talking about Premature Optimization, this behavior extends beyond that scope, and is often a symptom of Complexification.

There are several reasons to be wary of this temptation. Whatever the reason for trying the new algorithm, be it Premature Optimization or simply curiosity, the result is less readable and less maintainable code. This tradeoff would be acceptable if the algorithm fulfilled requirements that no simpler algorithm could, but this is often not the case. The algorithm's merits must be compared with its complexity to determine if it is overly complex for the situation at hand.

This leads us to another point that must be carefully considered. Does the algorithm provide the benefits that are implied by the algorithm's inventor? It is common for an algorithm to be made to sound much more useful than it is, particularly if the inventor spent a considerable amount of time researching it. Read carefully the description of the algorithm and sort out the verifiable claims as opposed to conjecture and vague praises. Once the benefits have been narrowed down and compared to similar algorithms for the given situation, the algorithm can be

---

ALL CODED UP WITH NOWHERE TO GO

Choosing an overly complex algorithm usually causes problems such as poor performance or maintenance difficulties, but in some cases, the results are even more disastrous. Just such a disaster happened and almost brought one project to its knees. The game programmer for the artificial intelligence (AI) chose a fuzzy logic system to use for controlling the non-player units. This complex AI system was supposed to adapt to the player and offer an appropriate challenge level for all players.

Instead, with only a few months left before the ship date, not a single non-player unit was moving. Plenty of code had been written for the complex AI module, but it was doing nothing useful. There was no hope that the fuzzy logic system would be done in time. However, defaulting to a much simpler system allowed the product to finish in time and still provided a formidable opponent for the player.

adopted if it provides a clear advantage. Even then, it is important to test the result of using the algorithm to verify that the expected benefits are in fact achieved.

Emergent Bugs

Excessive complexity can also arise in the interactions between code units. The most common causes of this are Premature Optimization, or the degradation of code due to changing requirements. Whatever causes it, this complexity can give rise to bugs that are difficult to reproduce and therefore difficult to find.

The reason for the difficulty in reproducing and identifying these problems lies with the concept of emergent behavior. Even a set of extremely simple systems can give rise to a complex set of interactions, as shown by Reynolds' boids [Reynolds87]. Reynolds simulated the complex behavior of a flock of birds, named boids in the simulation, by using a small set of simple rules to guide the behavior of each individual boid. When a group of boids was placed together, a complex flight pattern emerged similar to those seen in real flocks of birds. Although creating some form of emergent behavior is not very difficult, achieving desired behavior from an emergent system is much more difficult.

Thus, the more complex the interactions between the various code units become, the more likely that unwanted behavior will emerge in the form of bugs. Therefore, if a bug appears that is difficult to reproduce, one area to investigate is the complexity of the interactions around the area where the bug occurred. This can lead to finding a reproducible case, or at the very least show that the interactions need refactoring regardless of whether they are the cause of this particular bug.

PREVENTION

Preventing Complexification is about choosing the correct algorithm for the job. Let us look at some techniques that can help with this decision.

K.I.S.S.

When considering which algorithm to choose for a given situation, great weight should be given to the simplicity of the algorithm. This is known as the K.I.S.S. principle, or Keep It Simple, Stupid. By simple, we are referring to the readability and understandability of the algorithm. These qualities are directly proportional to the maintainability of the algorithm.

To better understand why this simplicity is important, consider the opposite direction. For each level of complexity added, either a group of programmers will be eliminated from maintaining that code or, at the very least, they will use more development time maintaining it. This does more than increase the development time for that algorithm alone. The programmers who are capable of maintaining the algorithm become a bottleneck for the team any time the algorithm must be changed. As more algorithms that are complex are introduced, the chance becomes greater that these bottleneck situations could result in the loss of development time for multiple programmers.

Know the Options

Before you can presume to choose the simplest algorithm, you must know what algorithms are available that fit your requirements. The good news is that many of the simpler algorithms are widely known. If you are considering an algorithm that is common knowledge, you increase the maintainability of the algorithm even if it is not the simplest. It is therefore important to understand what algorithms are well known within the computer science community. Consulting with the less experienced programmers, particularly those fresh out of college, is a good way to determine whether an algorithm is common knowledge.

When we talked about K.I.S.S., you were encouraged to choose the simplest algorithm. It might seem contradictory to now recommend choosing the most commonly known algorithm. This is not as much a contradiction as it might first appear, because algorithms that find wide adoption are generally those that are simplest to understand. This is a form of natural selection of algorithms, which supports the idea that simple algorithms are the most beneficial for software development. The longer the algorithm has been around to survive the selection process, the more likely it is to be simple and well known. However, as with natural selection, some algorithms that were once extremely effective and popular might have become obsolete with changes in the environment. In the case of software development, this environment is the hardware, which sees constant improvements, and inevitably has an effect on the software that is run on it. Although these algorithms will eventually disappear, it takes time for them to be removed by evolution, and there can sometimes be remnants that remain indefinitely.

There will be many times when a common knowledge algorithm does not suit the requirements of the given problem. If this is the case, the first step should be to reevaluate the requirements to determine if they can be relaxed to accommodate one of the simple and well-known algorithms. If the requirements cannot be relaxed to accommodate a common algorithm, it is time to do some research. Even if you know an algorithm that will work, it is a good idea to search for a couple of alternatives. With a selection of algorithms, make sure they each meet the requirements. Next, consider their readability, understandability, and maintainability. This should be the factor given the greatest weight in deciding which algorithm to use.

Start Small

This is a good point to step back a bit and consider the topic of requirements that we mentioned previously. The choice of algorithm is most constrained by the absolute requirements of the problem. Whereas the rest of the criteria for deciding which algorithm to use are relative and only affect the ordering; algorithms that do not meet the requirements are rejected completely.

Because of this rigidity, it is important to control the detail level of requirements so as not to eliminate too many functions. If the requirements are too strict, only complex functions are likely to be left to choose from, with the corresponding problems to long-term development that they bring with them. You should therefore strive to provide the smallest and least restrictive set of requirements possible during the initial phase of the project. The requirements can always be tightened as new information is provided or discovered. The resulting changes to the algorithms used are much easier because they will progress from a simpler algorithm to a more complex algorithm.

To understand why going from a simpler algorithm to a more complex algorithm is beneficial to development time, it is best to look at it from a risk management standpoint. Imagine that you start with the simpler algorithm, which is quicker to implement than the more complex algorithm. Imagine that a few minor modifications are required at first, which are performed in a short time due to the maintainability of the simple algorithm. Now one of two things could happen. The algorithm could be sufficient and ship with the application, or a more complex algorithm could replace the algorithm because of stricter requirements. Now take the opposite scenario by starting with the complex algorithm. This takes some time to implement, and then more time is used up when minor modifications are required. Because of its complexity, a senior programmer must be involved, which costs more development time. Once again, two possibilities arise. The application ships with the complex algorithm, or you discover that the simpler algorithm has a higher performance or allows certain operations that the complex one doesn't and therefore you must replace the complex algorithm with the simpler one. By comparing these two scenarios, you can see that on average the first scenario will cost less development time and therefore money. Even in the worst case, the development time is not much longer than the best case for the second scenario. Given that the first option has less risk, it is obviously the preferred choice.

The Right Language

While in theory all Turing complete languages, which contain both conditionals and loops, can execute any algorithm that can be executed in another Turing complete language, the reality of writing the algorithm varies across different languages. Language differences can make a difference in the apparent complexity of an algorithm, just as using external libraries can also reduce the apparent complexity.

To demonstrate the impact of the language used to implement an algorithm, let us look at two methods for adding two vectors. The first method uses standard C++ code:

```
// Desc:    Generic efficient vector template.
// Input:   i_size - Size of vector.
//          t_ValueType - Type of vector elements.
template <unsigned int i_size = 4,
typename t_ValueType = double>
class t_Vector
{
public:

        // Desc:    Provide access to value type.
        typedef t_ValueType t_Type;

        // Desc:    Provide access to size.
        static const unsigned int k_size;

        // ...

        // Desc:    Add vector element by element.
        // Input:   i_rhs - Vector to add to this vector.
        // Output:  Reference to self.
        t_Vector &operator+=(const t_Vector &i_rhs)
        {
            for(unsigned int l_index = 0;
                l_index < k_size; ++l_index) {
                m_value[l_index] +=
                    i_rhs.m_value[l_index];
            }
```

```
        }

private:

        // ...

        // Desc:    Array of values that form this vector.
        t_Type m_value[i_size];

};
```

The second method, which is accomplished through C++ template meta-programming, is much more complex:

```
// Desc:      Generic efficient vector template.
// Input:     i_size - Size of vector.
//            t_ValueType - Type of vector elements.
template <unsigned int i_size = 4,
typename t_ValueType = double>
class t_Vector
{
public:

        // Desc:    Provide access to value type.
        typedef t_ValueType t_Type;

        // Desc:    Provide access to size.
        static const unsigned int k_size;

        // ...

        // Desc:    Add vector element by element.
        // Input:   i_rhs - Vector to add to this vector.
        // Output:  Reference to self.
        t_Vector &operator+=(const t_Vector &i_rhs)
        {
            t_Add<i_size - 1>::m_Add(*this, i_rhs);
            return(*this);
        }

private:

        // ...
```

```
        // Desc:    Array of values that form this vector.
        t_Type m_value[i_size];

public:

        // ...

        // Desc:    Add elements at index and
        //          recurse index-1.
        template <unsigned int i_index> struct t_Add
        {
            // Input:    io_lhs - Vector to add
            //                     element from and into.
            //           i_rhs - Vector to add
            //                     element from.
            static inline void m_Add(
                t_Vector<i_size, t_Type> &io_lhs,
                const t_Vector<i_size, t_Type> &i_rhs)
            {
                io_lhs[i_index] += i_rhs[i_index];
                t_Add<i_index - 1>::m_Add(
                    io_lhs, i_rhs);
            }
        };

        // Desc:    Terminate recursion and add elements
        //          at index 0.
        template <> struct t_Add<0>
        {
            // Input:    io_lhs - Vector to add
            //                      element from and into.
            //           i_rhs - Vector to add
            //                      element from.
            static inline void m_Add(
                t_Vector<i_size, t_Type> &io_lhs,
                const t_Vector<i_size, t_Type> &i_rhs)
            {
                io_lhs[0] += i_rhs[0];
            }
        };

        // ...

    };
```

It accomplishes the same work as the first method, except it does it at compile time. In order to accomplish this, it can only use the limited programming environment for executing code at compile time. Despite the fact that this adds to the performance of the final application, using this special template language to accomplish this is only advisable if the extra performance is required. As a side note, see [Alexandrescu01] for more information on C++ template meta-programming.

While it is not always possible to choose the language in which the implementation will be done, when it is, consideration should be given to finding the language in which the algorithm can be expressed in the simplest most readable terms. Choosing different languages is becoming a more viable solution with the introduction of .NET, which simplifies the integration of multiple languages in one application. More detail on this is provided in Chapter 3, "NIH Syndrome," but the benefits apply here as well.

For example, some languages such as Perl support regular expressions. If you plan to write an algorithm for processing text, there is a good chance that it can be expressed in a more succinct manner using regular expressions. Therefore, it makes sense to use one of these languages when implementing this algorithm, or, at the very least, use a library that supports regular expressions. Depending on the circumstances, one choice might be preferable. While this can make the code less complex, it does only make it more readable if the programmer understands the language and syntax of the implementation. If you suspect that a programmer who does not understand the language and syntax might need to read the code, proper documentation and comments are essential. It can also be useful to point the reader to a good source for further information, such as [Friedl02] for regular expressions.

Try Explaining That

If you have any doubts that the algorithm chosen is too complex, it is a good idea to enlist the aid of an uninvolved third party. Start by explaining the problem and inquiring about possible algorithms that the other programmer believes would fit the situation. Do not reveal which algorithm you plan to use until after you have obtained the other programmer's suggestions. At this point, you might already discover that a simpler algorithm exists that you could be using.

If you are still convinced that you have chosen the right algorithm, proceed to explain the algorithm to another programmer. This will give you a better idea of the level of complexity of the algorithm, and in addition, will provide a guideline for the level of documentation that will be required for the algorithm to be maintainable. In this case, you should enlist the aid of a less experienced programmer if you want to get the best idea of how hard the algorithm is to understand.

Refactoring for Prevention

Most of the time, we talk about refactoring in the context of curing an illness, but refactoring is also a necessary part of the normal development process due to the normal decay of code quality caused by time and unavoidable changes in requirements. Refactoring can be particularly useful in preventing Complexification in the interaction between code units. Because the excess complexity arises as a byproduct of the normal development process, it is reasonable to expect to require the use of refactoring to keep it under control.

For most languages, reducing the complexity of interactions between code units through refactoring is accomplished by reducing the number of publicly accessible points of interaction on each encapsulated code units. Because the number of possible interactions increases substantially with each point of interaction available, removing even a few points of interaction can greatly reduce the complexity of an application. The primary means of reducing these public access points is through the movement of data and methods to more appropriate and less accessible locations in the code. Occasionally, code might turn out to be obsolete as well, allowing its removal altogether.

CURE

This illness is extremely common, and therefore you are likely to encounter its effects when working with other programmers' code. You might also encounter it in old code that you have written. First, let us look at when you should correct this problem, and then we can go into the proper techniques for refactoring.

When to Refactor

There are several reasons that can require the refactoring of a complex algorithm into a simpler algorithm. Let us look at each of them in turn. The first and most obvious reason for refactoring is when the algorithm is no longer meeting the requirements for which it was chosen. Algorithms that are more complex tend to also be more restrictive in their focus and therefore can easily become obsolete with changing requirements. When this occurs, it is also a good time to evaluate the requirements to see if they can be loosened to prevent future refactoring of the same sort from occurring. Regardless, there is no choice in this situation but to refactor.

Another extremely compelling reason to refactor a complex algorithm comes at the end of development in the optimization phase. The added complexity can often cost processing power. In many cases, requirements can be traded off for improved performance by replacing the complex algorithm with a simpler algorithm that requires less processing. As always, remember that this type of decision should only occur during the optimization phase to avoid Premature Optimization.

An opportunity for refactoring that is often overlooked occurs when the algorithm needs a minor modification. When confronted with the need to modify a complex algorithm, look at the requirements and context of the algorithm and determine if there is a prospect of simplifying the algorithm. If you do find that the algorithm can be simplified, this will not only make your modification easier, but also make any future modifications less time consuming. On average, this should work out to an increase in productivity that offsets the small amount of extra time spent in the short term.

Another useful reason for refactoring a complex algorithm is to come to a better understanding of the effects of the algorithm and how it operates. In this case, you might not end up simplifying the algorithm at all, but if not, you will, at that point, understand why it could not be simplified. You will also gain an understanding of the algorithm that can help if any modifications are required.

Simplify the Algorithm

The first step in refactoring a problem caused by Complexification is to choose a simpler algorithm to use. This process is much the same as we talked about in preventing Complexification, but there are a couple of extra considerations. Since an algorithm already exists, you must check to see if there is other code that depends on the algorithm in ways not specified by the requirements. In essence, this implicitly makes the requirements stricter. Before you can relax these requirements, you must learn what they are. Once you understand the unspecified requirements, you can then decide if they can be relaxed. In this case, if you want to relax them you must refactor the code to remove the dependencies before performing the refactoring to remove the Complexification.

Even if there are no implicit requirements, it is quite possible that the interface of the simpler algorithm will differ from the complex algorithm's interface. This refactoring should also be performed first. This ensures that the old algorithm's interface can be made to match to the new algorithm's interface. By doing this, it ensures that the algorithms will not alter the intended functionality.

Simplify the Code

Once you have implemented the new algorithm, it is important to also clean up the code. The first step should be to remove dead code from the new implementation. Dead code comes in two primary forms. The first and most obvious dead code is code that will no longer be executed under any circumstances with the new implementation. The second form of dead code is less obvious and has a much more detrimental effect on the readability of the code. This takes the form of code that is called but performs no work; in other words, it produces no change in the program state and no visible effect to the user. Examples of this form are temporary variables that are assigned but never used, and private functions with arguments that have the same value in all calls.

Code of both forms and any associated comments should be removed from the application completely. At this point, many argue that they might use the code again, or that they will need to refer to the code in the future. However, if you are using a source control system, this older code can be stored in earlier revisions without cluttering the present state of the code. If you are not using a source control system, you should be. At the very minimum, CVS is a freely available version control system that has been ported to many platforms. It is also useful to have a set of graphical tools, such as WinCVS for CVS, which allow you to browse the code history. With the use of a source control system, the reasons for simply commenting out the code or leaving it are greatly outweighed by the poor readability that the dead code causes.

The process of refactoring can also cause names and comments to become confusing, meaningless, or plain wrong. Each comment should be checked and updated if necessary, and any names that do not seem clear should be renamed. This might seem like a considerable amount of work, but doing it consistently will save you time when you make the inevitable changes that will need to be made.

RELATED ILLNESSES

Complexification is often closely tied to Premature Optimization. Optimizing an algorithm commonly results in a more complex algorithm. Thus, Premature Optimization becomes a motivating factor for choosing complex algorithms early in development. Vice versa, Complexification can introduce the symptoms of Premature Optimization even if the choice of algorithm was not based on optimization. This second connection is most often seen when adapting some nifty new algorithm just published in some programmer's favorite magazine. In the end, it does not matter so much which of the two illnesses that the problem is attributed to; both require the same refactoring steps to solve the problem. This is not to say that you can ignore one of the illnesses. They contain many areas that do not overlap, and each must be understood to handle those disparate areas. However, when they come together they are inseparable problems that are both solved at the same time.

Over Simplification is even more intricately intertwined with Complexification than Premature Optimization is. In the vast majority of cases of Over Simplification, one or more of the programmers involved will have encountered Complexification at least once but more likely multiple times. Over Simplification represents the wrong path to solving Complexification. When we talk about Complexification, the concept of choosing the minimal implementation is mentioned often. However, as with all the illnesses it is important to take into account the entire situation. The solution should be minimal, but it must also be complete. This is discussed in detail in Chapter 5, "Over Simplification," which we highly recommend that you read in conjunction with this illness.

As is true of most other illnesses, Complexification generally ignores the long-term effects of the decision to use a complex algorithm. This is the ever-present symptom of Myopia, which is a minor illness that runs through many of the other illnesses. Complexification exhibits Myopia in the problems it causes for future maintenance and in the wasted work that can occur if requirements change later in development. Remember to always consider the long-term effects of your decisions.

FIRST AID KIT

Very few tools help with Complexification, other than the usual refactoring tools and proper use of high-level languages. The challenge is learning to evaluate problems with the concept of avoiding unneeded complexity. Once you get into the proper frame of thought, very little work or assistance is necessary.

SUMMARY

The symptoms of Complexification are:

■   Poor readability might indicate overly complex code, but further investigation is necessary to determine if complex code is the root cause of the poor readability.

- The performance of extra computation that is undetectable to the end user.
- Use of an algorithm primarily because it was the most recently learned.
- The emergence of bugs caused by the overly complex interactions between objects or other code units.

To prevent Complexification:

- K.I.S.S., or Keep It Simple, Stupid
- Know or learn what algorithms will work for the problem.
- Start with the simplest or most well known algorithm, and only add complexity as necessary.
- If the option is available, choose the best programming language implementing the algorithm.
- Test how difficult it is to explain the algorithm, preferably to a colleague not working on the associated problem.
- Perform refactoring early and often to reduce the complexity of interactions between code units.

To cure the effects of Complexification:

- Do not be afraid to refactor to a less optimal but more general solution as long as the general solution still meets the requirements.
- Determine the least restrictive requirements that still solve the problem, and then choose the simplest algorithm that meets these requirements.
- Clean up code after simplifying to remove dead code and incorrect comments.

5 | Over Simplification

DESCRIPTION

When designing objects and interfaces, it is important to keep the design and implementation simple in order to maximize readability, usability, and maintainability. However, the goal of minimizing the complexity of the implementation can go too far, making the implementation essentially useless. Over Simplification comes from a desire to follow proper design techniques, but the programmer takes the wrong approach to this goal. Let us take a closer look at how to spot code that is too simple for the requirements placed upon it.

SYMPTOMS

Overly simple code is more difficult to identify that the overly complex code created by Complexification. This is true in part because complex code should be less common, and therefore easier to analyze than simple code. Since the majority of the code on a project should be simpler code, it becomes hard to identify the overly simple code from the correct level of simplicity. On the other hand, complex code tends to stick out and requires only an inspection to mark it as a candidate for being overly complex. So, what are the symptoms that we can use to help reveal overly simple code?

So Simple It Does Not Do Anything

The reason we write code is to accomplish some goal, or at least it should be. However, due to poor coding practices or the inevitable degradation of code over the course of a project, sections of code can cease to perform any useful function. Spotting this type of code serves two purposes. First, it improves performance by removing code that does not need to be called. Second, and more important, it makes the code more readable and maintainable.

What does this have to do with overly simple code? After all, you would not think of extra code coming from simplicity. To understand this, it is best to understand that simplicity does not always mean code simplicity. Simplicity can also mean design simplicity. While seeking design simplicity, it is easy to increase code complexity.

The creation of interfaces and class hierarchies is a common place for this to occur. In the interests of creating consistency in design and use of interfaces and objects, the programmer might be required to introduce code simply to handle the odd cases created by attempts to make all objects equal when they are not. This is generally caused by a misunderstanding of object-oriented technology. Although the design at first appears simple, in practice it becomes more complex and forces the addition of functionality just to maintain the simplicity of the design.

Shifting Complexity

Most applications contain layers of functionality built on top of each other. In a properly designed application, these layers should distribute the complexity as evenly as possible. By assigning the appropriate responsibilities to the appropriate layer, the implementation can be simplified while maintaining a simple interface to the other layers. However, if responsibilities are assigned to the wrong layers, the overall complexity of the code will rise.

One of the most common occurrences of this is Over Simplification in the lower-level functionality of the application. In an attempt to provide a simple base on which to build, the complexity is shifted upward where it becomes many times more complex to implement. This rise in complexity is caused by the differences in data access and the increased communication necessary between layers. This simplification is often accompanied by weakening of encapsulation and abstraction to allow the transfer of functionality to another location in the code. Whether this occurs through design failures or the natural evolution of the code, it is a sign of Over Simplification that should be corrected.

Do Not Touch

One roadblock to fixing Over Simplification can often be a sign of a programmer suffering from that very illness. Because there is a beauty in simplicity, a programmer

SIMPLY DISASTROUS

One project developed a very simple interface that was to handle the majority of the work for the application. It consisted of two main classes, and only one of these classes was designed for inheritance. Let us look at the basics of these two classes:

```cpp
class t_Object {

    // ...

private:

    // Linked list of object properties.
    t_Property *m_properties;

};

class t_Property {

    // ...

private:

    // Type identifier.
    int m_id;

    // Owner object.
    t_Object *m_owner;

    // Next property.
    t_Property *m_next;

};
```

Every instance of the main object would be the same except for the properties that were attached to it. Each property knew only about itself and its owning object. This design was chosen for its simplicity and consistency. In addition, several manager classes handled the execution of specific types of properties.

This simple system quickly broke down as it became obvious that properties were required to know about each other. Because the only way to distinguish the property was through its type, the code became scattered with small cut-and-paste instances of linked list searches for a particular property type. These were accompanied by type cast, introducing even more chances for error. Numerous bugs were encountered because of this originally simple design.

who creates a simple design and implementation might defend that implementation even if the change will result in increased simplicity for the code as a whole. This narrow vision is understandable from several perspectives. The time invested in writing the code and creating the simple design seems in danger, and modification to it will make the original programmer's job of maintaining it more difficult.

Nevertheless, these views ignore the necessary give and take required for a team to function efficiently. Making the changes will help in relations and also show a willingness to share in responsibility for the entire application. When you find yourself or another programmer resisting changes to functionality because of a fear of breaking the simplicity, step back and evaluate how the change will affect the simplicity of the code base as a whole. If you end up with a simple low-level implementation, but a horrible mess at the middle level and up, you will regret not moving some of the responsibilities down to the lower level.

PREVENTION

Over Simplification is prevented primarily in the design stages of development. Once the coding begins, the distribution of complexity is likely to take a life of its own and require redesign and refactoring to fix. This does not mean that there are not opportunities to prevent Over Simplification while coding, just that it is simply less common.

Completeness

When we talked about Complexification we emphasized the importance of a minimal interface and simple implementation, but that was only half of the equation. Completeness is also necessary to produce the best design and implementation. A minimal and complete implementation will require the least maintenance, and be the easiest to maintain.

The concepts of minimal and complete might seem at odds. This is not the case, but they do represent a balancing act. Added unnecessary functionality will make the application more than complete and therefore not minimal, whereas failing to provide necessary functionality will make the implementation more than minimal but fail to be complete.

Thus, a fine balance must be maintained in order to take advantage of the benefits of both simplicity and full functionality. Additionally, this balance might shift as changes occur to the design and other implementations in the application. This requires the flexibility to refactor the interface and implementation to restore the balance and continue to provide the best maintainability for the object.

Realistic Design

In order to create an implementation that is minimal and complete, the design that the implementation is meant to fulfill must be realistic. If you plan to write an application that does everything, you will end up with an overly complex implementation. Other programmers and perhaps even you will not be able to take advantage of the functionality included because it is obfuscated by the complexity of trying to understand its use. This usually means that you must aim for less than you would like to accomplish. In fact, you should generally aim for less than your first estimate of what you can accomplish.

On the opposite side of the design problem, do not try to get away with too simple of a design. An overly simple design and implementation will leave the missing functionality out, and this will not be discovered until later in the project. As with any changes later in the project, increasing the complexity of the design will result in a substantially higher cost.

Perhaps at this point you are still thinking that you do not want to end up implementing unused functionality. However, you must look at it from a more long-term perspective. A good analogy to draw upon is David Sklansky's discussion of winning at poker [Sklansky87]. Poker is about winning over the course of multiple hands, and even across an entire lifetime of playing poker. If you make a bet when the odds are in your favor, you have made money whether you win or lose that particular bet. If you make a bet when the odds are against you, you have lost money whether you win or lose that particular bet. The reasoning behind this is that by averaging the money won and lost across time, only betting when the odds are in your favor will result in a profit, whereas only betting when the odds are against you will result in a loss. Poker is not a game of individual hands, just as software development is not a matter of individual classes. While you might occasionally implement extra functionality at the beginning that loses some time, you have in reality won by saving the much higher late development costs in many other instances.

Winds of Change

The other reason for implementing functionality that is not of immediate use is to reduce the impact of likely changes. In software development, changes to requirements and therefore design are inevitable. However, some of these changes are more likely than others. By incorporating design elements that anticipate the most likely changes, you are once again betting when the odds are in your favor. Whether the particular change happens or not, you have made a winning choice.

Preparing for change is useful at any point during development, but it becomes of particularly importance toward the end of the project. At first, it might appear that anticipating change is most important at the beginning of the project. While it is true that anticipating possible changes early will save development time, unanticipated changes can be dealt with if there is still time remaining. As the project end nears, however, the impact of any change is magnified. If a change becomes necessary that is not easily made to the current code base, the project could be delayed, or even worse, canceled.

Determining what changes are likely is really a two-step process, starting with asking the customer what they expect will change. In this case, the customer does not necessarily represent the end user. It might be a manager to whom you are responsible, or another project that will be using your code. There is a definite benefit in reducing the number of people between the end user and the programmer, but having them talk directly to each other is not always feasible or productive. Depending on the project type and personalities involved, there might be a need for people in between.

Regardless, the second step must be to determine the possible changes the customer is not likely to consider. This usual involves technical issues, such as user interface considerations or development time. While these should be brought to the attention of the customer, they might still request that their approach be implemented. Therefore, the implementation should be designed to easily accommodate both the customer's expected changes and your expected changes.

CURE

Once Over Simplification has reared its ugly head within your code, it will continue to cause problems as long as it is there. Let us look at how to remedy the problem to save time and money.

More Sacred Code

Just as programmers defend the code they have written against replacement by outside code when suffering from NIH Syndrome, they also defend their code against added complexity when suffering from Over Simplification. Treating their code as sacred, they will defy reason in their desire to maintain its simplicity. Before you can consider refactoring, you must convince the owner of the code, which could very well be you, to consider the reasons why added complexity to the code might reduce the overall complexity of the entire code base.

As with NIH Syndrome, the defense of code simplicity might indeed be valid. Before ignoring the sacred code of the programmer, you must be disciplined enough to consider all consequences of the proposed change. The objective of a change that increases the complexity of a particular section of code should be to reduce the overall complexity of the entire code base. Resist the temptation to shift complexity to another programmer when it only benefits you and not the entire development effort.

When to Refactor

Over Simplification causes complexity to appear in the wrong areas of the code. As this complexity appears you should consider the benefits of refactoring, whether you are working on it at the time or just encounter it while browsing the code. It is possible to allow one instance of misplaced complexity to rest, if you are reasonably sure that the particular functionality will not be required again. As soon as similar functionality occurs more than once, the complexity should be refactored because the maintenance of the complexity increases substantially for each occurrence. Chances are high that more occurrences of that complexity will continue to occur.

Up and Down

Refactoring Over Simplification in most cases requires the movement of code between layers of functionality (Figure 5.1). In an object-oriented language, this might be represented by movement within a class hierarchy, but it might also involve movement between hierarchies. Do not constrain your movement of functionality by arbitrary language constructs. Over Simplification is a design problem and is not particular to any programming language features.

For example, if you find a function in one class that primarily uses access to members of another class, the function should be moved to the class with the members it is accessing. If you see:

```
class Object1 {

    // ...
```

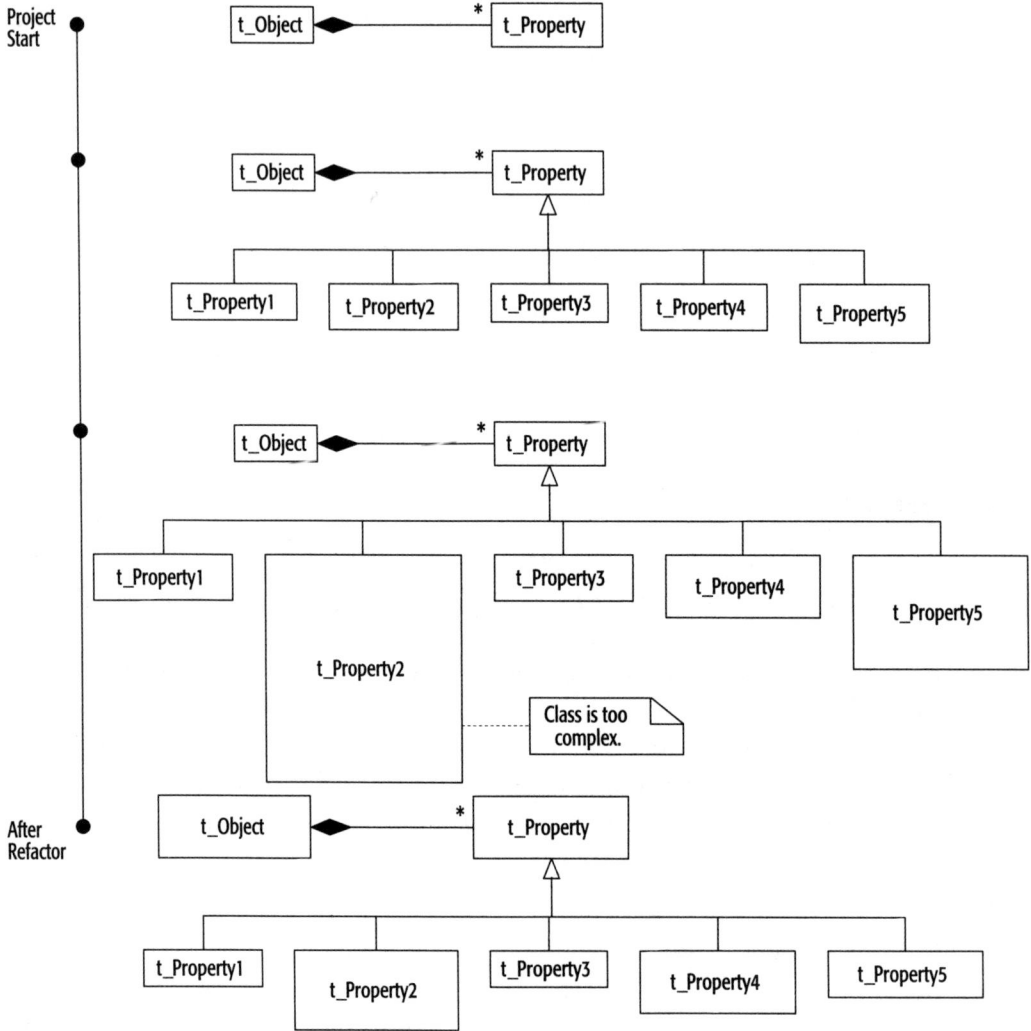

FIGURE 5.1 The progression of class complexity over a typical group of classes suffering from Over Simplification. The size of the rectangle representing a class represents the classes' complexity. Once it becomes obvious that one or more of the classes has become overly complex, refactoring can redistribute the complexity to a more maintainable level.

```
public void someFunction(Object2 object)
{
    int someValue = object.getSomeValue();
    int someOtherValue =
        object.getSomeOtherValue();
```

```
        // Do some stuff with someValue and
        // someOtherValue that does not use
        // any private members of Object1.

        object.setYetAnotherValue(result);
    }

    // ...

}
```

This should become:

```
class Object2 {

    // ...

    public void someRefactoredFunction()
    {
        // Do some stuff with someValue and
        // someOtherValue which are already
        // members of Object2.

        yetAnotherValue = result;
    }

    // ...

}
```

Duplicate code in multiple derived classes can also indicate an instance where the simplicity of the base class is being preserved at the cost of the derived classes. By moving these functions to a single function in the base class, the overall complexity of the class hierarchy will be reduced even though the base class complexity will rise. These, and other refactorings, should be done as soon as they are found to get the most benefit from the redistribution of complexity.

Libraries for Reuse

Future projects can also increase the benefits of refactoring. After your project discovers the proper level of functionality for a module, through experience and refactoring, that module can be separated into an independent library. This lasting benefit of refactoring is often overlooked because of production deadlines and the excuse that you will never use the code again anyway. To be competitive in the modern world of software development, however, rewriting everything from scratch is no longer a reasonable option. Managers are demanding more code reuse, and you could be stuck with old code despite your desire to throw it away. Since refactoring does have benefits for your project and it will save you headaches if you must reuse the code, it is only logical to maintain the simplicity and reusability of the code without concern for future assumptions.

RELATED ILLNESSES

Previous contact with Complexification is the primary cause for Over Simplification. After having been on one or more projects that suffered from overly complex design and code, it is easy to switch to the other extreme and attempt to make the design and code overly simple. Instead, a middle ground should be sought to balance the complexity of individual sections of the code with the complexity of the code base as a whole. By seeking to attain minimal and complete interfaces and implementation, you will not be susceptible to either Over Simplification or Complexification.

Just as Complexification causes Over Simplification, Over Simplification is a common cause of Brittle Bones. Because Over Simplification offers functionality that is generally less than complete, it provides a weak foundation for future code built upon it. This makes it important to remedy Over Simplification early in development. Failure to do so is symptomatic of the common illness of Myopia.

FIRST AID KIT

As was the case with Complexification, very few tools can help with Over Simplification. Collecting the proper level of requirements and covering the necessary design issues will put you well on your way to preventing Over Simplification, and for this, you primarily need knowledge, which is not something that a tool can give you.

SUMMARY

The symptoms of Over Simplification are:

- Code that performs no useful function, either by design or through decay, and that is still present.
- Complexity that has been shifted from its correct location and therefore been made more complex.
- Excessive protectiveness placed on code just to maintain its simplicity.

To prevent Over Simplification:

- Ensure that the design, interfaces, and implementation are complete in addition to minimal.
- Be realistic and do not make the low-level design too simple; otherwise, you will be missing functionality that is required for the higher-level design.
- Prepare for areas of likely change.

To cure the effects of Over Simplification:

- Do not allow code to become a sacred cow because of its simplicity.
- Add or move complexity when refactoring as necessary, rather than to avoid work or responsibility.
- Make sure that functionality is at the proper place and level to minimize complexity.
- Consider the possibility that the code will be reused when writing it.

6 | Docuphobia

DESCRIPTION

Poor documentation is a common contributor to bugs and a hindrance to code reuse. While perfect documentation is a practical impossibility, much of the current documentation does not even reach the rating of passable. The largest problem is not bad documentation, but no documentation at all. Many programmers have an aversion to writing prose; they might have even taken up programming to avoid it in the first place. Nevertheless, the fact remains that documentation is important to the software development process, particularly when working in a team environment. This documentation does not have to be Shakespeare, but it must be written to help others understand your code.

SYMPTOMS

The most obvious sign of Docuphobia is the lack of documentation. However, even if documentation exists, it does not mean it is good documentation. There are a few main ways to decide whether you feel the documentation is sufficient.

What, How, and Why?

Documentation should answer three main questions for the end user of the code. This user might be another programmer, or it might be you after you have forgotten what you did. No matter who the end user is, the documentation must be clear enough to make the code usable and to avoid any potentially common errors.

The first question that will interest the user is: what does this code do? The user should know what inputs the code accepts and what output can be expected. The user also needs to know what side effects, or global changes, can be expected and what errors might occur. All this information is necessary if the user is to use the functionality properly and handle any potential problems. As a user, you can only spot missing documentation if you attempt to use the function in a way that is improper but not documented as such. You could also analyze the workings of the functionality, but this is unacceptably time consuming.

For instance, you encounter the following function declaration:

```
/** Interpolate linearly between two numbers.
 *      @param      i_start - first number
 *      @param      i_end - second number
 *      @param      i_percentage -
 *                              percentage to interpolate
 *      @return     Interpolated number.
 */
float linear_interpolate(float i_start,
        float i_end, float i_percentage);
```

You might then try using it:

```
linear_interpolate(5.0f, 15.0f, 50.0f);
```

However, this causes the program to report a failed assertion. The cause for this is obvious once the function implementation is revealed:

```
float linear_interpolate(float i_start,
    float i_end, float i_percentage)
{
    assert(!(i_percentage < 0.0f) &&
        !(i_percentage > 1.0f));
    return(((i_end - i_start) *
        i_percentage) + i_start);
}
```

This error could have been avoided with proper documentation:

```
/**     Compute the value along the linear slope
 *      between two numbers.
 *      @param      i_start - number at 0%(0.0f)
 *                      on the slope
 *      @param      i_end - number at 100%(1.0f)
 *                      on the slope
```

```
 *      @param     i_percentage - percentage along
 *                 slope at which value occurs
 *                 as a decimal factor from
 *                 (0.0f,1.0f) inclusive.
 *      @return    Interpolated value between
 *                 (i_start,i_end) inclusive.
 */
float linear_interpolate(float i_start,
     float i_end, float i_percentage);
```

The second question is: how does this code accomplish its goal? Documentation for this question does not need to be made public unless it affects the use of the functionality, but it still should be done. Documenting the internals eases maintenance and provides information for refactoring the code if part of it can be used elsewhere. If an error occurs and the documentation is lacking, either the original programmer must dredge up the memories of how it operates or some programmer must examine the code and puzzle out what it does.

This small snippet of Ruby code illustrates the use of comments for explaining how an algorithm works:

```ruby
# Quadratic equation is
# (-b +/- sqrt(b^2 - 4ac)) / 2a
# of which the discriminate is the part inside
# the square root. The discriminate is used to
# determine the number of solutions
# the equation has: 0, 1, or 2.
discriminate = (b * b) - (4 * a * c)

# Positive discriminate means 2 solutions.
if discriminate > 0 then
        solution =
            [(-b + Math.sqrt(discriminate)) / (2 * a),
             (-b - Math.sqrt(discriminate)) / (2 * a)]
# Negative discriminate means no solutions.
elsif discriminate < 0 then
        solution = nil
# Zero discriminate means 1 solution.
else
        solution = -b / (2 * a)
end
```

This provides the necessary extra information to remind the programmer what the quadratic equation is and how the discriminate is used. Having this information readily available with the code prevents other programmers from having to put the equation back together and guess at the purpose of the discriminate, as they would have to if the code read like this instead:

```
discriminate = (b * b) - (4 * a * c)
if discriminate > 0 then
        solution =
            [(-b + Math.sqrt(discriminate)) / (2 * a),
            (-b - Math.sqrt(discriminate)) / (2 * a)]
elsif discriminate < 0 then
        solution = nil
else
        solution = -b / (2 * a)
end
```

The final question is: why does it do it that way? This is the most often overlooked task of documentation. Without the background behind why it was done a certain way, programmers can accidentally modify code in ways that break the functionality without knowing it. The other reason that this question is extremely important to document is that it cannot be extracted from the implementation itself. If necessary, both what and how can be determined by analyzing the code even without documentation. However, the code cannot reveal the motivations of the programmer or programmers behind its inception.

Take this single line of code:

```
m_nextUniqueID += 2;
```

It is easy to see that this line adds two to the next unique identifier, but why use 2? The surrounding code offers little help:

```
if(!i_instance) {
    return(t_Handle(0, 0));
}

unsigned int l_freeIndex = m_freeIndex;

if(l_freeIndex >= m_handles.size()) {
    m_handles.resize(l_freeIndex + 1);
}
```

```
if(unsigned int l_nextFreeIndex =
    m_handles[l_freeIndex].m_GetHandle().m_GetIndex(
    ) {
            m_freeIndex = l_nextFreeIndex;
} else {
    ++m_freeIndex;
}

m_handles[l_freeIndex] =
    t_HandleInstance<t_Type>(l_freeIndex,
    m_nextUniqueID, i_instance);

m_nextUniqueID += 2;

return(m_handles[l_freeIndex].m_GetHandle());
```

However, if the original programmer had added the appropriate comment, the reasoning would have been obvious:

```
// Increment the unique identifier by two, since
// it is initialized to one it will never be zero
// even if it wraps around.  While this does not
// guarantee an absolutely unique identifier, the
// chances are miniscule that both index and
// identifier will be duplicated in any
// reasonable amount of time.
m_nextUniqueID += 2;
```

Documentation is sufficient once it has answered these three questions. For the end user, this can only be discovered through trial and error. Therefore, it is important for the original programmer to make the best effort to properly document. Reviewing the documentation can help in this process, and should be done on a regular basis. Additionally, having other programmers review the documentation and ask questions can point out areas that lack complete information. If you are already doing code reviews, this should become a part of them. If you are not doing code reviews, you might want to consider starting them.

Not That Way

When functionality is used incorrectly, there are two main explanations. The first explanation is that the programmer who attempted to use the functionality did not read the accompanying documentation. If you are such a programmer, try making a new habit of reading the documentation. Otherwise, you can expect to encounter many problems for which you will only have yourself to blame.

The second and equally common explanation is that the documentation was not clear or sufficient to prevent the code from being used incorrectly. If you find someone using your code incorrectly, check your documentation. See if it can be extended or made clearer. Be sure to ask the programmer who used it incorrectly if the new documentation explains the situation better. If you are the end user and you discover a problem, take it to the original programmer and ask that it be updated. If the original programmer is no longer available, update it to the best of your knowledge.

For a good example of function misuse, we can look back on the example from the last section where we talked about how comments should explain what the function does:

```
/** Interpolate linearly between two numbers.
 *     @param    i_start - first number
 *     @param    i_end - second number
 *     @param    i_percentage —
 *                     percentage to interpolate
 *     @return    Interpolated number.
 */
float linear_interpolate(float i_start,
    float i_end, float i_percentage);
```

If you wrote this function, and the accompanying comments, you might later come across this usage of your function:

```
linear_interpolate(5.0f, 15.0f, 50.0f);
```

You realize this will not work, because you assume the percentage is between 0.0f and 1.0f. Chances are that if you ask the programmer who wrote this why he chose the number 50.0f, he will tell you that he wanted 50 percent and assumed that the percentage value should be between 0.0f and 100.0f. The programmer should have instead used 0.5f for the percentage argument. Thus, the comments obviously need more detail.

PREVENTION

Preventing poor documentation is about discipline and having the correct set of tools and practices. The discipline is up to you, but here we can cover the techniques and tools that can make proper documentation easier to accomplish.

Too Little vs. Too Much

Before we discuss documentation techniques in detail, it is important to confront a common explanation that is given by programmers suffering from Docuphobia. They will claim that too many comments make the code confusing and more difficult to read. Therefore, they feel justified in providing very little documentation. In truth, providing excessive documentation is not optimal, but it is better than providing no documentation. Although the extra documentation requires extra time to wade through, it will often offer information that is inaccessible from the code alone. If there was no documentation, there might be no recourse for discovering the needed information. The best approach to documentation is to aim for the same goal as code design: minimal and complete. However, in cases of doubt it is better to err on the side of completeness.

Document in the Code

The majority of the documentation should be written with the code to which it is associated. This provides a locality of reference that makes it easy to find and easy to maintain. This removes a common barrier to proper documentation, which is the natural aversion to interrupt the workflow to find and fill in documentation located in some other location. There are two main approaches to documenting within the code, and you might hear many arguments about the benefits of one approach over another. Perhaps the best approach is a combination of the two approaches. Let us look at each of these approaches and discuss how they can be used together.

Self-Documenting Code

An integral approach to code documentation is to write the code so that it documents itself. In other words, the code should be naturally readable in such a way as to make it easy to understand. This is accomplished by proper naming conventions and statement formatting. The advantages of this are clear, since the code must be written anyway; it is only a minute amount of extra work to make it more readable.

For example, take this Java class:

```java
class Bomb {

    Bomb(int ticksLeftToDetonation) {
        this.ticksLeftToDetonation =
            ticksLeftToDetonation;
    }

    boolean updateTime(int ticks) {
        assert(ticks > 0);

        if(isTriggered()) {
            explode();
        } else {
            countdown(ticks);
        }

        return isTriggered();
    }

    private boolean isTriggered() {
        return(ticksLeftToDetonation <= 0);
    }

    private void explode() {
        System.out.println("BOOM!");
    }

    private void countdown(int ticks) {
        ticksLeftToDetonation -= ticks;
    }

    private int ticksLeftToDetonation;

}
```

In particular, let us focus on the updateTime method. First, notice that the name of the method tells us that we are going to be updating the time. Next, the name of the argument, ticks, gives us an indication that we are providing the number of ticks that we want to add to the time. We could have made this clearer, such as ticksToAdd, but since the meaning was relatively clear without the extra wording, we chose not to make the name longer.

Next, we have the `assert` call, which makes it evident that we expect `ticks` to be greater than 0. This is a language feature for checking that the state of the program is as expected. This type of functionality might be removed from the final release build for performance reasons, but in the meantime, it provides another means of detecting problems early. It also provides information to programmers reading the code about the expectation the writer had about the state of the application, and any inputs and outputs. This type of language feature is available in many languages, some of which have more than one variation to further improve the self-documenting nature of the language.

Now we want to know if the bomb is triggered, which can be easily understood from the statement `if(isTriggered())`. This could be made slightly clearer by actually providing the implicit `this` variable, but since almost all programmers understand that it is implied, it can be seen as redundant. Thus, `if(this.isTriggered())` is slightly clearer, but not enough to warrant the extra characters. Additionally, it might cause some programmers to think there is a reason for the distinction when there is none. Leaving `this` out avoids this misleading idea. Finally, the function names `explode` and `countdown` are made to be straightforward indications of what to expect from these function calls. The function `explode` causes the object to explode, in the example we simply print `BOOM!`, and the function countdown advances the bomb's `countdown` timer.

Self-documenting code is useful for understanding the implementation as it is written, but it cannot help if the implementation is not available for perusal. This requires separate documentation that describes the usage of the functionality without providing implementation details. This limits the use of self-documenting code to the names available in the interface to the functionality. In most cases, this is insufficient, but we can still keep the documentation as close to the code as possible by putting in the comments. Later, the documentation can be automatically extracted into a different form if necessary.

Comments

While self-documenting code should be the goal when writing the code itself, it cannot provide all the information necessary to understand the code. This is where code comments, which do not affect the final application, come in. Comments are particularly important for answering the question of why. They are also useful for clarifying the implementation of sections of code that cannot be made self-documenting due to the constraints of the language.

The following is an excerpt from a handle manager class used to generate unique handles for associating with data pointers. This code, available in full on the companion CD-ROM in Source/Examples/Chapter6/why.h, shows several examples of comments that explain why a particular code statement was written:

```
// Desc:    Handle manager for a particular class.
// Input:   t_Type - Type of object for which the
//          handles are associated.
template <typename t_Type> class t_HandleManager
{
public:

    // Desc:    Initialize manager.
    // Notes:   The unique identifier is
    //          initialized to one and
    //          incremented by two in order
    //          to ensure zero is never a
    //          valid unique ID.
    t_HandleManager()
        : m_nextUniqueID(1), m_freeIndex(0) {}

    // Desc:    Create a new handle and
    //          associate it with a pointer.
    // Input:   i_instance - Pointer to object
    //          instance that handle will
    //          represent.
    // Output:  Unique handle that represents
    //          the provided pointer.
    // Notes:   This manager does not prevent
    //          two handles from representing
    //          the same pointer.  If this is
    //          the case make sure that all
    //          handles associated with the
    //          object are destroyed when the
    //          object is deleted.
    t_Handle m_Create(t_Type *i_instance)
    {
        if(!i_instance) {
            // A handle with a unique ID of
            // zero is always invalid.
            return(t_Handle(0, 0));
        }

        // Make a local copy of the
        // current free index so the stored
```

```
        // free index can be updated.
        unsigned int l_freeIndex = m_freeIndex;

        if(l_freeIndex >= m_handles.size()) {
            m_handles.resize(l_freeIndex + 1);
        }

        // If the free index points to
        // a handle that has a valid index
        // then the stored index should
        // be set to that index.  This
        // happens because handle destruction
        // fills in this field when
        // if cleans out the handle.
        if(unsigned int l_nextFreeIndex =
            m_handles[l_freeIndex
            ].m_GetHandle().m_GetIndex()) {
                m_freeIndex =
                    l_nextFreeIndex;
        } else {
            ++m_freeIndex;
        }

        m_handles[l_freeIndex] =
            t_HandleInstance<t_Type>(
            l_freeIndex, m_nextUniqueID,
            i_instance);

        // Increment the unique identifier
        // by two, since it is initialized
        // to one it will never be zero even
        // if it wraps around.  While this
        // does not guarantee an absolutely
        // unique identifier, the chances
        // are miniscule that both index and
        // identifier will be duplicated
        // in any reasonable amount of time.
        m_nextUniqueID += 2;

        return(m_handles[l_freeIndex
            ].m_GetHandle());
    }

    // ...

}
```

There are several comments in this code snippet that explain why a particular approach was used, and several other lines are not commented because they are relatively easy to determine their purpose. For example, when we add two to obtain the next unique identifier it is not immediately obvious why two is important. If this were not commented on, another programmer might change the two to a one. This would violate the expectations of the class and could cause strange errors to occur. Therefore, we point out that the use of two eliminates zero as a valid identifier assuming the first unique identifier is one. This important assumption is also stated when the first unique identifier is initialized to prevent another programmer from changing that value and once again breaking the assumptions that the class makes. This line of reasoning also applies to the comment about creating an invalid identifier. This is a case, however, where the code could be made self-documenting by creating a static instance of an invalid identifier in the handle class, and giving it the name `INVALID_HANDLE`.

Also, notice that simpler operations, such as creating and returning the handle, are not commented. There is little reason to comment these lines of code because they are straightforward, and other programmers can be expected to change or maintain them without much risk of causing problems elsewhere. Another example of such a straightforward line of code is the incrementing of the free index. While it is better to comment too much than too little, striving to achieve the optimal balance should be a continuing effort that will make the code clearer and the important comments more evident.

Another important use of comments is to preserve encapsulation and abstraction. This might sound odd knowing that comments do not affect the language usage directly. However, without comments there would be no way for the end user of an interface to know what functionality there is and what the proper inputs are without examining the implementation. By examining the implementation, the interface's user can easily produce too tight of a coupling to the implementation details. This makes future changes to the implementation difficult. While some of the usage information can be transferred with good naming conventions, it is not possible to present a sufficient amount of information in that form. Therefore, the interface must be documented with comments for it to be used properly.

For example, here is the interface to a configuration class designed to read a configuration file and provide simple configuration querying:

```
/**     Stores configuration file information
 *      for later retrieval.
 */
class configuration
```

```
    {
    public:

        /**
         *      Read the configuration file and
         *      store information in class.
         *
         *      @param     i_filename name of
         *                         configuration file.
         */
        explicit configuration(
            const std::string &i_filename);

        /**
         *      Check if a named configuration value
         *      exists in this configuration.
         *
         *      @param      i_name name to lookup.
         *      @return     true if name is in configuration,
         *                  false if it does not exist.
         */
        bool has(const std::string &i_name) const;

        /**
         *      Retrieve named configuration value
         *      from this configuration.
         *
         *      @param      i_name name to lookup.
         *      @return     Copy of string value
         *                  associated with name,
         *                  or empty string if named
         *                  value does not exist in
         *                  this configuration.
         */
        const string find(
            const std::string &i_name) const;

    };
```

This code is also available on the companion CD-ROM in Source/Examples/ Chapter6/interface.h. If you read the description of the constructor and two methods for this class, you should notice that no implementation details are provided. This allows hiding of the implementation so that future changes can be made without worrying that existing code will break because of its reliance on the internal implementation. The documentation does provide the necessary information to pass correct information into the methods and to interpret the information that is received back from some of the methods. This eliminates the need for the user of the class to look at the implementation even if it is provided, and gives the developer the option to provide only the interface and binary libraries for the implementation.

Coding Standards

A set of coding standards, which include rules on documentation, can assist in both self-documenting code and proper commenting. With a set of rules to follow, it is easier to maintain the necessary discipline for good commenting. It also provides a consistency in the code and documentation that makes it easier to read and search.

For example, the following are some rules that would be beneficial to have in the coding standards for a project:

- All public documentation must be formatted according to the JavaDoc specification.
- All publicly accessible member functions, including those accessible to derived classes, must have their declaration fully documented.
- Do not provide implementation details in the documentation except when it directly affects the use of the function.
- All private member functions should be documented as completely as possible.
- All argument constraints must be provided in the function's documentation.
- All side effects must be documented as a warning.
- Any exceptions thrown by the function should be documented to the best of the developer's knowledge.

This example represents a portion of the rules that could be provided as part of a coding standard. Since it is important that the programmer remember and continue to use these rules, the list should not be made excessively long. Only the rules for common occurrences should be presented here; rules for less common situations can be discussed and decided upon in code reviews and personal correspondences.

Avoiding Duplication

Duplicating information for the purposes of documentation is both tedious and error prone. It also leads to difficulties in maintaining the documentation. This can be demoralizing to the point of discouraging the programmer from writing any documentation. As we will discuss shortly, the majority of this duplication is unnecessary with the modern tools available.

For example, suppose you have the following Java function, also found on the companion CD-ROM in Source/JavaExamples/com/crm/ppt/examples/chapter6/ DuplicationExample.java.

ON THE CD

```
/**
 * public float computeAverage(float values[]);
 *
 * Compute the average of an array of values.
 *
 * @param     values array of float values that
 *            must not be null or zero length
 * @return    floating point sum of values
 *            divided by length
 */
public float computeAverage(float values[]) {
    // Set sum to zero.
    float sum = 0.0f;

    // Start at index zero and increment
    // it by one until the index is not
    // less than the length of the
    // values array.
    for(int index = 0;
        index < values.length; ++index) {
        // Add the value at the current index
        // to the sum.
        sum += values[index];
    }

    // Return the sum divided by the length
    // of the values array.
    return(sum / values.length);
}
```

First, notice that the function name and signature are repeated in the interface documentation comment. Perhaps this was done automatically by a template created for your editor, or maybe you used cut-and-paste to copy the code line into the comment. Either way, this is not a good practice to follow. If another programmer comes along to change the function, he must also modify the comment. This is tedious and error prone, often leading to just ignoring the comment update and thus making the resulting documentation incorrect.

Notice also the commenting within the body of the function. These comments are only saying what is happening, which can already be easily deduced from the code. If even an operator is changed, the comment must also be updated or become incorrect. This can become really confusing during debugging as the programmer attempting to correct the code cannot determine if the original programmer meant to perform the operation in the code or the operation in the comment.

Instead of this redundant commenting, what the commenting should look like is:

```
/** Compute the average of an array of values. */
public float /** Average. */
    computeAverage_(
    float values[] /** Must not be null or
                            zero length. */
    )
{
    float sum = 0.0f;

    for(int index = 0;
        index < values.length; ++index) {

        sum += values[index];
    }

    return (sum / values.length);
}
```

You can see here that the code is used as part of the documentation, particularly the variable names. Maintaining a style of commenting that uses the code as part of the documentation reduces ambiguity and maintenance errors, as well as saving the redundant cut-and-paste or typing of the information the first time around. Unfortunately, the tools currently available do not support this style if you want to generate external documentation from the code and comments. There is still a need to repeat the argument names in the interface comments. However, much of the documentation can be derived directly from the code itself and paired with the comments to form complete external documentation with the help of a documentation generator. This means that with modern tools you can lean more toward the second style of commenting than the first.

Automation

Several steps in the documentation process should be automated. In fact, you should strive to automate as much of it as possible to encourage proper documentation, and focus the effort on the portion that cannot be automated because the information only exists in the programmer's mind.

First, automate the extraction of documentation from the comments contained within the code. The majority of languages have one or more applications that can accomplish this task. For example, Java has JavaDoc as a standard documentation tool, and the freely available Doxygen serves a similar purpose for C++. Both of these, along with many of the freely available tools, require the comments to follow a format for documentation to be extracted. If you have a large code base that is not commented in any of these styles, there are commercially available applications that can still extract decent documentation from the existing comments.

The true benefit of these applications is their understanding of the language that is being documented. This allows the generator to create browsing and search information not only based on the comments but also on the structure of the code itself. Most of these tools also offer several different output forms, among which HTML is the most useful (Figure 6.1). By adding the documentation generation to your build process, the resulting documentation can be placed on a server accessible to the team and updated with every build. This gives each member of the team ready access to the documentation through a Web browser that is available on almost every computer these days.

Another automation step that should be added to the build is documentation error checking. While the conceptual aspect of the documentation cannot be verified, semantic errors such as misnamed or missing parameters can be checked. The resulting list of errors should be added to the available documentation and, if possible, mailed to the code owner for timely correction.

On a more individual level, automating the process of writing the comments that will form the documentation can be done with most modern editors. Some of them contain a certain built-in amount of knowledge for creating comments associated with language constructs. Whether or not this functionality exists in your editor of choice, you can still take advantage of code creation features and auto-

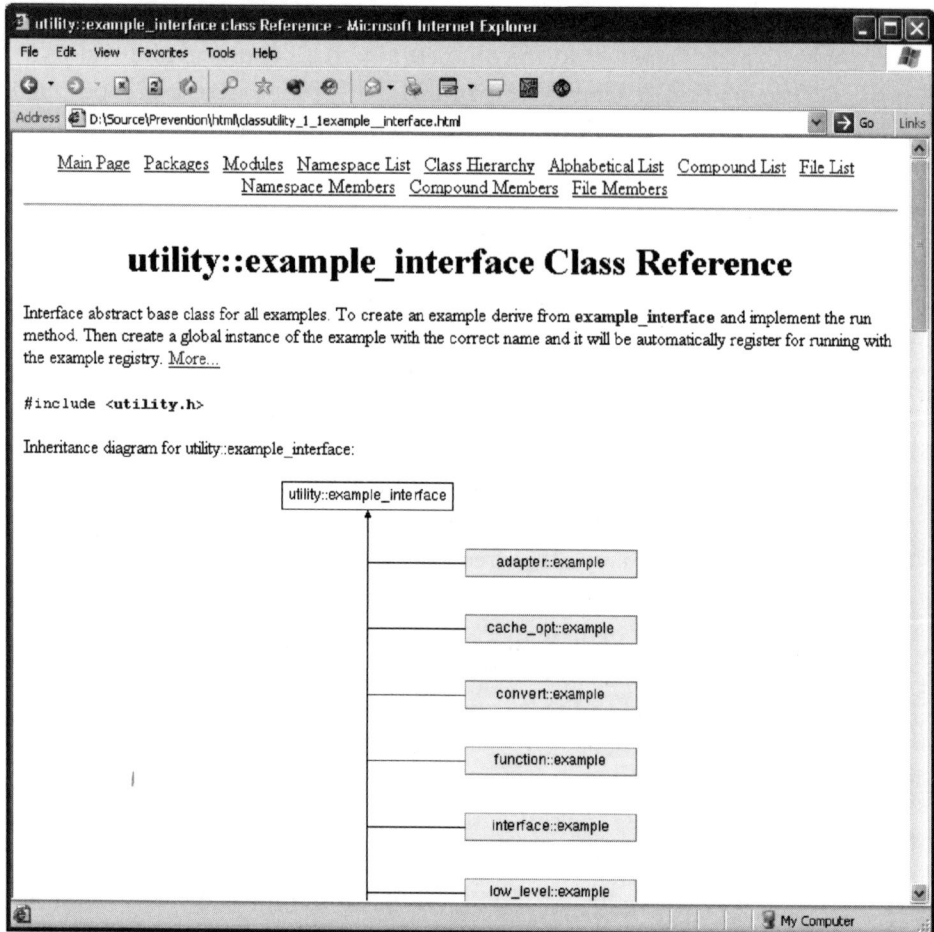

FIGURE 6.1   Example of HTML documentation generated by Doxygen from the code and associated comments.

completion to create your own shortcuts for generating the structure of the documentation comments. This leaves you with only the unique information to fill in for each new comment.

Maintenance

For the documentation to be useful, it must be correct. A common failing is for the documentation to be correct at the beginning of the project when the code is first written, only to degrade as functionality is changed and refactored without the necessary updates to the documentation. With every code change, you must check the corresponding documentation to determine if it requires updating. If so, it must be done immediately to avoid the problems of out-of-sync code and documentation. This is where providing the majority of your documentation within the code itself helps tremendously. With the documentation close by, it is much more likely to be updated to maintain consistency than if another application must be launched and more files searched.

CURE

Unless you are abnormally lucky, you will encounter reams of old code that is not properly documented in your career. Much of this code is useful despite the lack of proper documentation. However, this code will continue to prove enigmatic to each new programmer if you do not take the necessary steps to document it.

Analyze

The first step to documenting undocumented code is to determine what documentation is missing. The most important documentation to provide is interface documentation. This is necessary for other programmers to understand the conventions for calling the foreign code. Other helpful documentation is good but not nearly as important. There are automated tools that can be used to assist in this process, but often the documentation is missing altogether and easy to spot even without automated assistance. Other documentation errors are harder to spot without automated assistance, so if you have access to automated checking it should be used. For example, a single argument missing from the documentation of a function with six arguments would only be spotted on a complete reading of that documentation along with examining the function declaration.

Incorrect documentation is different from missing documentation. This is difficult to detect automatically, and can often only be determined through the course of using the functionality. Therefore, it is not possible to analyze the code initially to find this form of missing documentation. When it is encountered during development, the documentation should be updated as soon as possible to reflect the correct use and result of the functionality. The one positive aspect of this problem, however, is that the functionality is in use and therefore it is more readily apparent what it does.

Query

Never hesitate to ask the original developer of the code for information to help document that code. This can often be the best source of information you will find for explaining undocumented code. However, this is a fallible source of information, as the programmer's memory might fail. To reduce the chance of getting incorrect information, you should collect available information about the code so you can ask directed questions. More focused questions will help narrow the search range and jog the programmer's memory. Be cautious to always ask a question without providing an answer or suggestion to prevent filling a memory gap with a falsely created memory from your suggestion.

Refactoring

There are two important relations between documentation and refactoring. First, consider the refactoring of older code that you did not write. As you refactor, you will gain a better understanding of the code. This gives you the information necessary to fill in missing documentation. This documentation will preserve the work that you put into learning what the code was doing for safe refactoring. Show discipline and write the documentation as soon as you discover what is happening in the code. This will preserve the most accurate understanding of the code without the possibility of memory errors.

The second and perhaps more important consideration when refactoring is to properly update the comments associated with the code that is being refactored. Actually, it should work both ways. Before refactoring, read the documentation that is associated with the code that is to be refactored. With this knowledge, you can ensure that the refactoring will not affect the code that uses the functionality as long as they only require the function to meet its documented specifications. This is a reasonable assumption since the internal implementation should never be used as an indicator of future functionality. If the behavior must be modified to meet new requirements or fix a bug, be sure to change the documentation once the changes have been made. You must also be sure that the documentation of any usages is up to date, and follow this to the first unaffected location in the call hierarchy. This is not a lot of extra work, as you must follow this call hierarchy anyway while doing the refactoring.

RELATED ILLNESSES

Even in well meaning programmers who started out providing proper documentation, Docuphobia can occur after too much contact with poor documentation. One of the causes of poor documentation is lack of maintenance associated with the CAP Epidemic. As code is cut-and-pasted, programmers often forget to update the comments to match the new context. This results in incorrect documentation that can mislead other programmers and cause subtle errors to occur. Nevertheless, the solution is not to avoid the documentation, but to learn how to write and maintain it properly. Avoiding cut-and-paste when possible is one of the steps toward this goal.

A more direct relationship is shared between Docuphobia and i. Proper naming skills and conventions lead to self-documenting code. This reduces the need for comments and therefore reduces the duplication that must occur if a comment is needed to explain a poor name. Proper naming is an important step toward proper documentation, and as was the case with avoiding cut-and-paste, helps avoid poor documentation that can lead to Docuphobia.

Finally, we come to the illness that is the cause of so many other illnesses. Skipping documentation appears to save time, but this is only an illusion suffered from Myopia. The time spent fixing miscommunications and relearning the purpose of each piece of code will easily dwarf the minor savings of skipping documentation. Thus, the long-term savings is achieved by proper documentation.

FIRST AID KIT

Most of the tools for documentation center on either self-documentation of code or the extraction of documentation from the code and comments automatically. Let us look at each of these types of tools separately.

Self-documentation is possible without any extra tools to some degree in all programming languages; simply choose the best names from code elements.

Nevertheless, some programming languages have extensive support for creating self-documenting code.

One language feature that is geared directly toward self-documenting code is Design by Contract programming. This is an integral part of the Eiffel programming language. The idea is to provide a visible contract that a function must fulfill. This ensures that both the user of the function and the writer of the function understand the necessary preconditions, post-conditions, and invariants that the function must obey. Because this is part of the language, it is enforced by the compiler and is not prone to the same synchronization problems that plague code comments. Other languages have similar extensions, such as iContract for Java, but portability is reduced because it is not an inherent part of the language.

It is also possible to create your own domain-specific language that uses terminology directly from the target domain. While this is appealing, a considerable amount of work is involved. Loss of portability and a high learning curve can also be deterrents, although the learning curve is reduced if the language is well designed and the programmers must learn the language of the target domain anyway. Nevertheless, if you decide to pursue this course, the most commonly used tools are lex and yacc, which can be used to create custom compilers and interpreters from a standard language specification. See [Levine92] for more detail.

Automated documentation systems come in a couple of varieties. Several freely available systems, such as JavaDoc and Doxygen (Figure 6.2), require a particular format to the comments that they parse. As long as this is begun at the beginning of the project, this is an excellent method for providing appropriate comments as well as automatic generation of separate documentation. Some editors, such as IntelliJ's IDEA, are also able to use these for providing tips and information while editing (Figure 6.3).

The second type of automated documentation is useful if you have a legacy system or one later in development that does not follow any documentation standards. Applications such as Doc-O-Matic from toolsfactory are able to extract reasonably accurate approximations from existing comments based on various heuristics.

Another tool that is very useful for improving automated documentation is documentation validation, such as iDoc for Java. These tools check to ensure that the comments match the code at a syntactic level, catching some of the more common oversights caused when code is changed but the comments are not updated. Documentation validation should be integrated with the build process for documentation to ensure that the latest documentation accurately reflects the way the

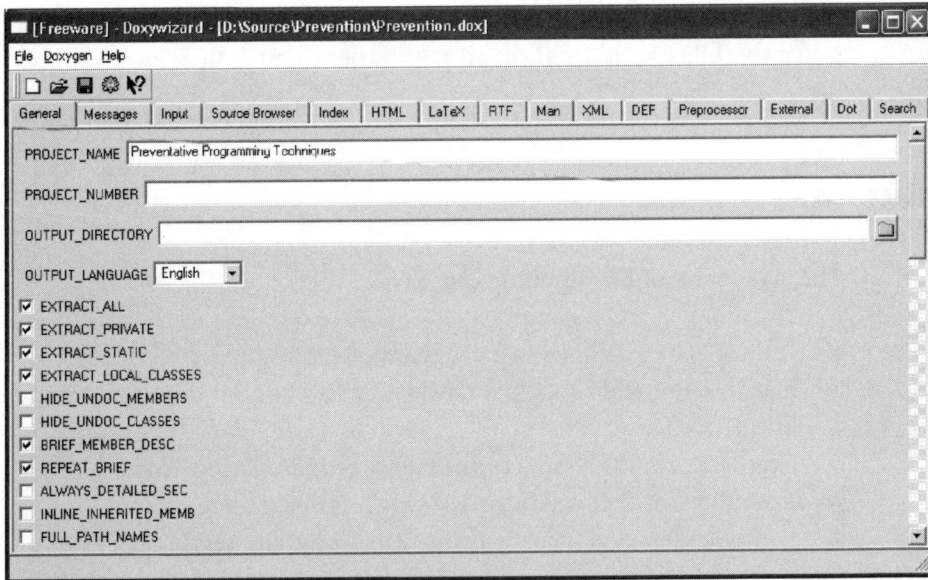

FIGURE 6.2   Configuration dialog for Doxygen, an automated documentation generator. Notice that several different output formats can be seen in the tabs, such as HTML, LaTeX, RTF, and man pages.

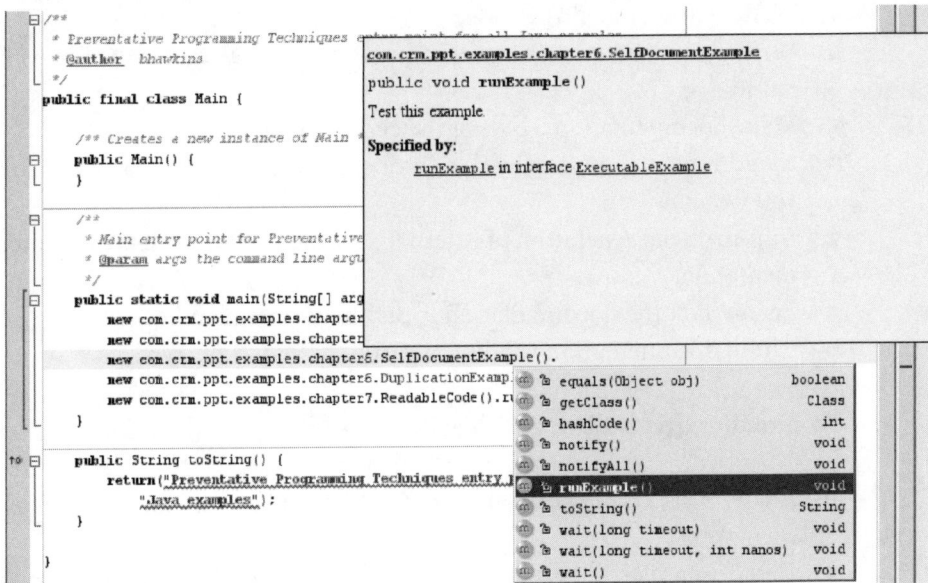

FIGURE 6.3   A tool-tip appears in IntelliJ's IDEA to show the JavaDoc comments for a member function.

code works. Some editors, such as IntelliJ's IDEA, might also provide this functionality while you edit, making it even easier to catch these oversights.

SUMMARY

The symptoms of Docuphobia are:

- A total lack of comments located anywhere in the code is the most obvious.
- Header comments that do not explain what the code does and, when necessary, how it does it.
- Comments in the source code that are only repeating what is already apparent from the source code, instead of why it is being done.
- Source code called incorrectly by other programmers.

To prevent Docuphobia:

- Find a balance that provides enough comments to make understandable code without becoming a tedious chore that does not add extra meaning.
- Write self-documenting code.
- Write documentation in the comments of the code rather than in some external source.
- Make documentation a part of the coding standards.
- Avoid information duplication in documentation as much as possible just as you would in code.
- Automate the generation of external documentation from the comment documentation.
- Ensure that the documentation is easy to maintain and update.
- Check documentation validity as part of the build process.

To cure the effects of Docuphobia:

- Analyze the current documentation, looking for missing or incorrect documentation.
- Determine what the proper documentation is based on the code and questions to the original code writer if possible.

- Ensure that you fully understand what the code is doing before documenting it, or request that someone who already knows the code document it.
- Add missing documentation as part of the learning process that is inherent in refactoring.
- Update documentation as you refactor.

7 i

DESCRIPTION

The title of this illness might have you wondering. This is exactly the same response programmers have to poorly named variables. Without a meaningful name, the only way to determine what a variable does is to examine its usage or read its documentation, if any is available. Even worse, if the variable is an undocumented member of a library to which you do not have the source code, the only possibility is through trial and error usage that might never fully reveal what the variable does. Therefore, the name of the illness itself should serve as a simple reminder to always provide meaningful names within the code.

The particular choice of i for the name comes from one of the most common constructs in the C language, the loop iteration:

```
for(int i = 0; i < end; ++i) {
    // Do something with i...
}
```

Here the i stands for iteration. Original memory was sparse even for the source code, and in the case of interpreted languages, the source code might have been the program itself. Because of this, and the desire to type less, programmers chose short names that often had only an obscure connection to their meaning or no connection at all. However, despite the abundance of memory available and the tools for writing code faster, this tradition still hangs on in many cases.

SYMPTOMS

What constitutes a poor name? The simple answer is that the name should explain exactly what is does or represents. However, this can lead to excessively long names that are cumbersome to read. The real answer is that the name must provide enough information so that its usage can be determined without much effort by the average programmer. Even this answer does not provide enough information for generating good names while programming. This answer serves only as a guideline by which to judge the names that are created. In the end, if the name serves its purpose well, it can be considered a success.

Comment Required?

One indicator that the name is not sufficient for its purposes is that a comment is required to explain what the variable is even when the context in which it is used is available. Of course, this criterion does not apply to names that are made public without their corresponding implementation details. These require commenting, because there is no context available. However, you will find many cases where the function name is not sufficient even if the implementation is available for parsing.

Suppose you encountered a section of code that looked like this.

```
if(itd()) {
    // ...
}
```

What exactly does itd mean? Perhaps it could be made clearer by adding a comment:

```
// If this object is initialized...
if(itd()) {
    // ...
}
```

However, if you need the comment, what purpose does the name X serve? You end up with one set of text for the compiler to read and another for the programmer to read. This could easily get unsynchronized and cause numerous errors. It also makes the code harder to read. Instead, we could choose a better function name and let it communicate to both the compiler and the programmers:

```
if(isInitialized()) {
    // ...
}
```

This function name typically needs no explanation or comment for another programmer to understand that it is meant to check the initialization state of the object to which it belongs. A short comment might be required, however, for some automatic documentation systems. In this case, it is generally kept extremely short and to the point since it is not necessary except for external documentation.

Nevertheless, there can be cases where even a straightforward function name such as this might hide important information. Imagine that to be initialized, several functions must first be called other than just a function with the name `initialize`. This information will quite often be important to the caller of `isInitialized`. Since the name would become excessively long if these conditions were to be self-documenting, the use of a comment is called for here.

This leads us to the observation that when the comment is conveying information to the programmer that is already conveyed to the compiler in a less readable form, the names used are insufficient and therefore wasteful. Comments related to the implementation details should only be used to explain why, and when the algorithm is complex, sometimes how the code operates. The comments should not be necessary to explain what the code elements mean. Look for comments that are simply restating what the code does with different names, especially if the names used in the code sound ambiguous. This is a clear symptom of poorly chosen names.

Of course, if you don't comment at all you will never discover any of this. Always read the code you have just written to determine if it is understandable. Pretend that you do not already know what it does and imagine what might be missing if you came at it from a fresh perspective. This is not easy, and even the best programmers will end up with code that is not fully understandable. An additional step that can help substantially is to have another programmer review the code to see if he understands what it does and why. Regular code reviews are a good means to provide this feedback without interrupting the normal workflow. Additionally, pair programming is another good way to obtain constant feedback from another programmer. This latter approach is an important part of the Extreme Programming methodology.

Dangers of Context

One danger to be particularly aware of, because of its misleading nature, is the context of the variable or code element. It is best to choose names that are still understandable despite small changes in the context. To better understand what we mean, we return to the name of this illness and its seminal example:

```
for(int i = 2; i < 512; i *= i) {
    bitMask &= i;
}
```

Notice that i might be understandable within its current context because of the small scope of its existence. However, code does not stay fixed over time. For example, the loop contents might grow substantially over time, causing the initialization of i to be lost at the top, and the meaning then requires a search. Another possibility is that i will be moved outside the loop:

```
bitMask &= i;
```

Once again, the meaning might require a search to make clear what i does and how it should be used. In particular, it is not clear that i represents a bit-field with a single bit set. Actually, even in the first code snippet it is unclear what i is ultimately meant to represent, only that it is a power of two and therefore represents a bit-field with a single bit set. Therefore, although it might seem safe at the time, do not rely too heavily on context to provide the full meaning of a variable or code element.

Variables named according to their context can also end up with misleading names in anther context when the actual meaning of the variable changes. For example, you could be using an interpolation algorithm in one place to interpolate position, with the corresponding variable name position. If this same code snippet was then changed to interpolate color, but the variable name was not changed, it would be confusing to other programmers what the real intention was of the code. Instead, the variable should be renamed to color. Of course, if you are cut-and-pasting the code you should instead consider creating an interpolation function with a generic name such as value.

PREVENTION

Discipline is the main virtue required to prevent this illness. Learn to always consider the name you give every variable, method, class, and so on. If you do this consistently, you will be well on your way to writing more understandable code. There are, however, some concerns that must be answered and some tips to help you maintain this discipline.

Editors: Tools of the Trade

Clear names need to be of a reasonable length, and this is where the most resistance is encountered to adopting good naming schemes. Except for rare special circumstances, memory usage is no longer a concern when it comes to name length. The primary argument that remains, therefore, is that typing longer names is time consuming. This argument fails in the face of ever-advancing editor technology. When we talked about Premature Optimization of the source code, we introduced many of the technologies that remove much of the typing from long variable names. Auto-completion stands out as the primary improvement that eliminates much of this typing. As discussed, this context-sensitive technology enables the programmer to type only a few characters before the rest of the name can be automatically completed by the editor (Figure 7.1). Learning to take advantage of this technology is important to improving code readability while maintaining efficiency.

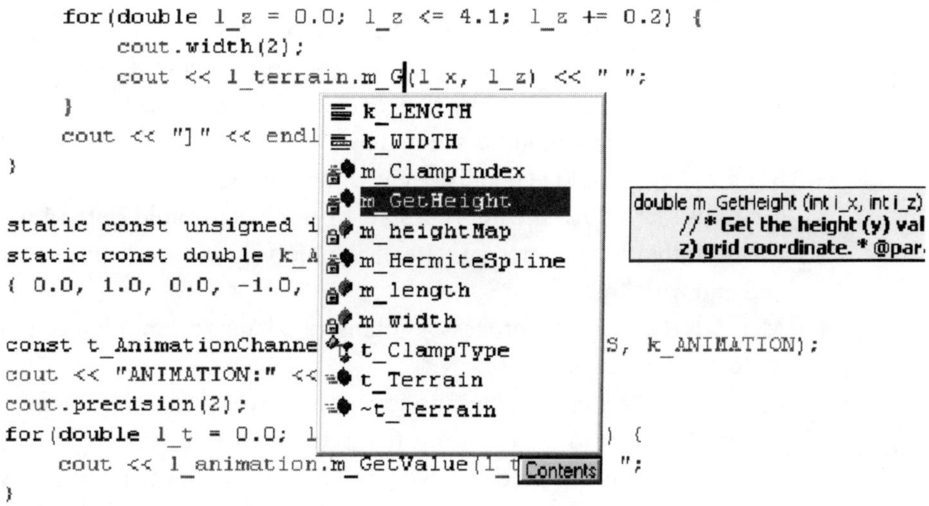

```
for(double l_z = 0.0; l_z <= 4.1; l_z += 0.2) {
    cout.width(2);
    cout << l_terrain.m_G|(l_x, l_z) << " ";
}
cout << "]" << endl
}

static const unsigned i
static const double k_A
{ 0.0, 1.0, 0.0, -1.0,

const t_AnimationChanne                    S, k_ANIMATION);
cout << "ANIMATION:" <<
cout.precision(2);
for(double l_t = 0.0; l                    ) {
    cout << l_animation.m_GetValue(l_         ";
}
```

Completion list entries:
- k_LENGTH
- k_WIDTH
- m_ClampIndex
- **m_GetHeight**
- m_heightMap
- m_HermiteSpline
- m_length
- m_width
- t_ClampType
- t_Terrain
- ~t_Terrain

Contents

Tooltip:
```
double m_GetHeight (int i_x, int i_z)
    // * Get the height (y) val
    z) grid coordinate. * @par.
```

FIGURE 7.1 Improved auto-completion with Visual Assist makes it easier to enter long names in Microsoft Visual Studio.

Some claim that these longer names make the code less readable rather than more readable, but this is rarely the case in practice. It is true that the names can become too verbose, which is a danger that must be avoided as well. However, this is rarely the main problem, as most programmers err on the side of names that are too short and unclear. While we should strive to achieve balance between length and understandability, it is better to err on the side of understandability. A name that is too long can be refactored by anyone to be a better length, but a name that is too short is much more difficult to lengthen because the meaning might not be obvious.

Naming Conventions

Establishing naming conventions is important to providing consistent names across a project. These conventions also assist in maintaining discipline because there is a permanent reference upon which to base name choices. In some cases, such as Java, the language community will have adopted certain standard naming conventions. This is even more beneficial as it allows code from different developers to share some common features that make reading easier.

- Class names should always start with a capital letter, and each subsequent new word should be capitalized as well.
- Constants should be all uppercase letters, separating words with underscores.
- All other methods, variables, and other named constructs should begin with a lowercase word with all subsequent new words capitalized.
- Method names should follow the general form of a verb followed by a noun, with other possible modifiers added in.
- Boolean tests should generally begin with is or has.
- Methods that return the internal state of an object should start with get.
- Methods that set the internal state of an object should start with set or change and end with To.
- Methods that perform processor-intensive calculation should begin with compute or calculate.
- Fields should generally be nouns, except in the case of boolean flags, which follow rules similar to the functions that retrieve these values.
- Arguments should be prefixed with io_ if they are input and output values, o_ if they are output values, or no prefix if they are input only values.
- Local variables should be prefixed with l_ if they have the same name as an argument.
- Template arguments should begin with t_.

Many of the conventions in this example are specific to the C++ language, but all languages can benefit from good naming conventions. Always check to see if naming conventions exist, and follow them. This will benefit those who read your code in the future, and assist you in the creation of names, which can always be a difficult process.

There is one common rule that should be followed across all languages. Just as you would do when designing interfaces, strive to make your names minimal and complete. This rule follows directly from the fact that the names are part of the interface to a class and are therefore automatically part of the minimal and complete goal. A complete name is one that requires little or no explanation to determine what action it will perform or what it represents. A minimal name is one that requires minimal space, allowing for easier reading by eliminating the need to parse an entire diatribe for just one name.

Everyday Language

We have offered some suggestions on why the variables should be of sufficient length and how to maintain the discipline needed to write good names, but what do we do to create better names? A good name should make clear what the code element represents in as succinct a form as possible. Naming conventions might also provide some guidance on what name to choose.

Perhaps the best method for choosing a good name, however, is to base the choice on how the resulting name will read in the code. This means that any programmer who is reading the code should be able to parse it as easily as he would normal everyday text such as you would find in a book or newspaper. While it is not possible to write text exactly as you would in your normal language, the choice of good names can make the code easily translatable into that normal language as the reader parses it.

The following Java snippet, also available on the companion CD-ROM in Source/JavaExamples/com/crm/ppt/examples/chapter7/ReadableCode.java, illustrates this technique:

ON THE CD

```
//...

/** Interpolation key string. */
private static final String
    INTERPOLATION = "INTERPOLATION";

/** Linear interpolation string value. */
private static final String
    LINEAR = "LINEAR";

/** Spline interpolation string value. */
private static final String
    SPLINE = "SPLINE";

/** Current configuration. */
private final Configuration
    configuration = new Configuration();
```

```
/** Configuration stored as
    key/value string pairs. */
private class Configuration {

    /** Has the configuration been
        fully filled in. */
    private boolean isInitialized;

    /** Store key/value pairs in a
        hash map for easy lookup. */
    private final HashMap
        map = new HashMap();

    /** Proxy class to make value
        access more readable. */
    public class Value {

        /** Internally stored string value. */
        private String value;

        /**
         * Check if value matches string.
         *
         * @param value string to compare value to
         * @return true if string match
         *         according to equals, false if
         *         null or comparison is false
         */
        public boolean is(String value) {
            if(value == null) {
                    return (false);
            }
            return (value.equals(this.value));
        }

        /**
         * Set the string of this value.
         *
         * @param value new string
         */
        public void to(String value) {
            this.value = value;
```

```
        }

    }

    /**
     * Change the initialization state of
     * this configuration.
     *
     * @param isInitialized true means the
     *          configuration is initialized,
     *          false means it is not initialized
     */
    public void changeInitializationFlagTo(
        boolean isInitialized) {
        this.isInitialized = isInitialized;
    }

    /**
     * Check to see if configuration is initialized.
     *
     * @return true if configuration is
     *          ready, false otherwise
     */
    public boolean isInitialized() {
            return (isInitialized);
    }

    /**
     * Check to see if a key exists in the
     * configuration.
     *
     * @param key string value of key to lookup
     * @return true if key exists, false if
     *          key does not
     */
    public boolean has(String key) {
        return (map.containsKey(key));
    }

    /**
     * Get the value of a key in order to
     * change its value. If the key/value pair
     * does not exist it is created.
```

```
     *
     * @param key string value of key to set
     * @return value to set or change string of
     */
    public Value setValueOf(String key) {
        if(!map.containsKey(key)) {
            map.put(key, new Value());
        }
        return (Value)(map.get(key));
    }

    /**
     * Get the value of a key, if the
     * key/value pair does not exist
     * the value represents a null String.
     *
     * @param key string value of key to get
     * @return value representing string
     */
    public Value getValueOf(String key) {
        if(!map.containsKey(key)) {
            return (new Value());
        }
        return (Value)(map.get(key));
    }

}

/**
 * Initialize the configuration for this example.
 */
public void initialize() {
        configuration.setValueOf(
            INTERPOLATION).to(LINEAR);
        configuration.changeInitializationFlagTo(true);
}

/**
 * Draw key frame function designed to
 * illustrate readable code naming.
 */
public void drawKeyFrame() {
    if(configuration.isInitialized() &&
```

```
        configuration.has(INTERPOLATION)) {
        if(configuration.getValueOf(
            INTERPOLATION).is(LINEAR)) {
            drawThisAt(
                resultOfLinearCalculation());
        } else if(configuration.getValueOf(
            INTERPOLATION).is(SPLINE)) {
            drawThisAt(
                resultOfSplineCalculation());
        }
    }
}
```

First, look at the initialization function, which can be read as:

For the configuration, set the value of interpolation to linear. Also, for the configuration, change the initialization flag to true.

Notice that we only had to fill in a few words to get this to read as ordinary language. This was accomplished mostly through appropriate naming, with a little help from a proxy class to allow the word to to be placed in the correct location. This proxy class allows us to provide much more readable code with minimal overhead. Notice also that it can provide extra error checking and debugging possibilities if necessary.

Now look at the drawing function, which can be read as:

If the configuration is initialized and has an interpolation value, then, if the configuration interpolation is linear, draw this object at the result of a linear calculation; otherwise, if the configuration interpolation is spline, draw this object at the result of a spline calculation.

Notice that this also only requires the addition of a few added words to be read correctly and easily. Because the words and semantic structure are similar to everyday language, code reading can be done much faster. Even more important, the code can also be understood more quickly, because of the natural language processing capabilities of our brain that we have been developing for much longer than our knowledge of any programming language.

CURE

As with many of the other illnesses, numerous examples of code that exhibits this illness exist. Some of this code you are likely to want to use or be required to use. As you use this code, you can refactor the names used to improve the readability. This will not only help other programmers on your team, but also help you when you need to change or fix the code later in development.

Read the Code

As you incorporate code into your project that is not hidden away in a library, read the code to come to a better understanding of what the code is doing. As you read the code, you might discover that certain names do not read naturally. These names slow the parsing of your code and are the ones that require refactoring.

Once encountered, you must first be sure that you understand what is occurring in the code. If you are sure you know what is happening, then proceed to replace all instances of the variable or code element with a new name. Refactoring and code examination tools can help ensure that all instances and only the correct instances are replaced (Figure 7.2). Once the new name is in place, reread the code to determine if the readability has increased. If not, choose a new name and repeat this procedure.

Comments to Code

The naming process can be assisted if the code has been commented because of the poor name choices that were used. If you encounter comments that are explaining what is happening, then you should attempt to incorporate these comments into the code itself. This process helps accomplish several improvements to the resulting code. The code is easier to read, both because the naming has improved and the comments are no longer in the way. Further, the comments represented a duplication of information that required extra maintenance.

To give a better idea of this process, suppose the drawKeyFrame code from earlier in this chapter started out looking like this:

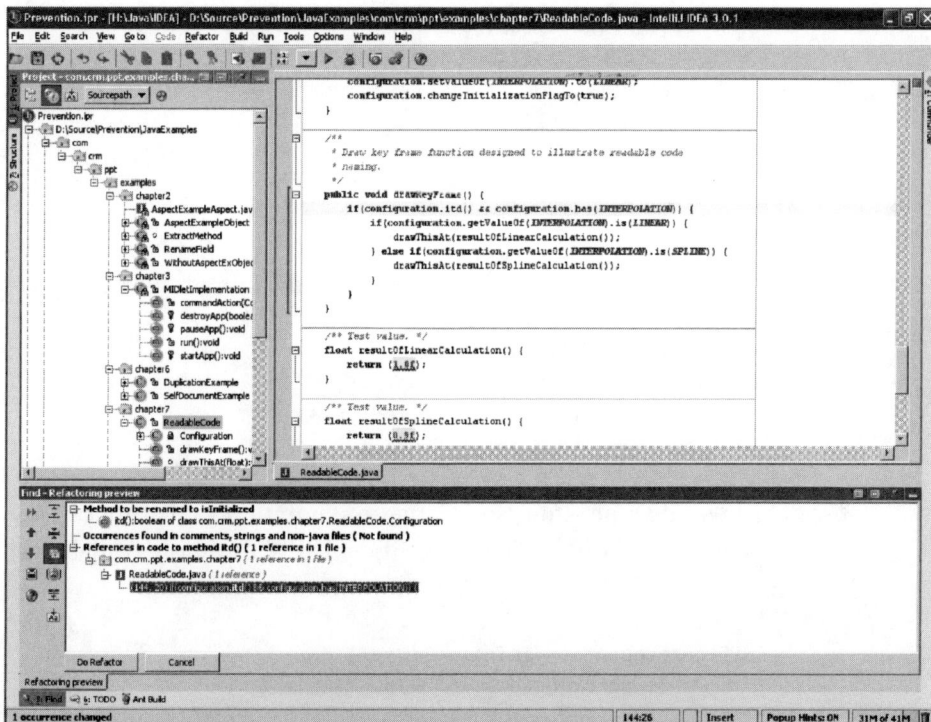

FIGURE 7.2 Refactoring a member function name using IntelliJ's IDEA to make it more readable. IDEA ensures that all instances are properly renamed.

```
/**
 * Draw key frame function designed to
 * illustrate readable code naming.
 */
public void drawKeyFrame() {
    // If the configuration is initialized and has
    // a value for interpolation...
    if(configuration.itd() &&
        configuration.get(INTERP) != NONE) {
        // If the configuration value for
        // interpolation is linear, draw this
        // object at the value calculated using
        // linear interpolation.
        // Otherwise, the value should be spline
        // and so we draw this at the value
        // calculated using spline interpolation.
```

```
            if(configuration.get(INTERP) == LIN) {
                draw(lin());
            } else {
                draw(spl());
            }
        }
    }
```

First, we remove the need for commenting the initialization check:

```
    // If the configuration has
    // a value for interpolation...
    if(configuration.isInitialized() &&
        configuration.get(INTERP) != NONE) {
```

Then we add a new function to the configuration so that we can just check if the value exists without having to get it, which is also more readable with a new name:

```
    if(configuration.isInitialized() &&
        configuration.has(INTERPOLATION)) {
```

Now we change the equality into a function call and give the retrieval function a better name:

```
            if(configuration.getValueOf(
                INTERPOLATION).is(LINEAR)) {
                // Draw this
                // object at the value calculated using
                // linear interpolation.
                draw(lin());
            } else {
                // Otherwise, the value should be
                // spline and so we draw this at
                // the value calculated using
                // spline interpolation.
                draw(spl());
            }
```

Next, we make the spline check more explicit, both for readability and to avoid problems if other types are added to the configuration file:

```
if(configuration.getValueOf(
    INTERPOLATION).is(LINEAR)) {
    // Draw this
    // object at the value calculated using
    // linear interpolation.
    draw(lin());
} else if(configuration.getValueOf(
    INTERPOLATION).is(SPLINE)) {
    // Draw this at
    // the value calculated using
    // spline interpolation.
    draw(spl());
}
```

We then change the name of the draw function to indicate what is being drawn:

```
if(configuration.getValueOf(
    INTERPOLATION).is(LINEAR)) {
    // Value calculated using
    // linear interpolation.
    drawThisAt(lin());
} else if(configuration.getValueOf(
    INTERPOLATION).is(SPLINE)) {
    // Value calculated using
    // spline interpolation.
    drawThisAt(spl());
}
```

The interpolation functions also need better names to indicate what they are doing:

```
if(configuration.getValueOf(
    INTERPOLATION).is(LINEAR)) {
    drawThisAt(
        resultOfLinearCalculation());
} else if(configuration.getValueOf(
    INTERPOLATION).is(SPLINE)) {
    drawThisAt(
        resultOfSplineCalculation());
}
```

Finally, we put it all together to get a more readable function:

```
/**
 * Draw key frame function designed to
 * illustrate readable code naming.
 */
public void drawKeyFrame() {
    if(configuration.isInitialized() &&
        configuration.has(INTERPOLATION)) {
        if(configuration.getValueOf(
            INTERPOLATION).is(LINEAR)) {
            drawThisAt(
                resultOfLinearCalculation());
        } else if(configuration.getValueOf(
            INTERPOLATION).is(SPLINE)) {
            drawThisAt(
                resultOfSplineCalculation());
        }
    }
}
```

RELATED ILLNESSES

Docuphobia and i are closely related due to their impact on the understandability of the code. Once the i illness is conquered, Docuphobia is easier to remedy because less extra documentation is necessary. However, good naming does not fully replace the need for comments and other forms of documentation. This is particularly true of interfaces where the programmer should be discouraged from looking at the implementation. Good naming can still be applied to the interface names, but extra documentation should be made available to lessen the temptation to look at the implementation. Thus, preventing these illnesses is normally a joint effort.

Another close relation between Docuphobia and i is one of the common reasons they both occur. This is where Premature Optimization comes into play; only this time it is theoretical optimization of development time saved by short names and no documentation. Experience, unfortunately, tells a different story. Without the proper readability to the code, more time is lost fixing bugs that should have never happened, and parsing code that should not have to be looked at.

The failure of short names to provide savings in development time is a result of Myopia. As is all too common, shortcuts taken early can lead to a much longer road later. Spend the extra time up front to provide good variable names and you will make it back when you do not have to do extra work at the end of the project.

FIRST AID KIT

The most important tool to assist in proper naming is a good editor. Look for features such as auto-completion and code templates that increase the speed of code entry. IntelliJ's IDEA has excellent auto-completion for Java, and the Visual Assist plug-in for Microsoft Visual Studio offers similar quality auto-completion and code templates for C and C++.

The only other tool that has a major impact on naming and code readability is the choice of programming language. Every language can be made more readable, but for certain tasks, a particular language might be more appropriate. If the option to choose the language is there, take it, but if other factors make that impossible, it is usually not difficult to make the chosen language read easier.

SUMMARY

The symptoms of i are:

- Code that requires redundant comments to make it clear what the code is doing.
- Code that cannot be easily parsed in one pass similar to reading ordinary language.
- A name that is not understandable without the surrounding context.

To prevent i:

- Use auto-completion and code templates available in most major editors to make writing fully understandable names easier.
- Follow standard naming conventions.
- Choose names that make the code read closer to everyday language.

To cure the effects of i:

- Read existing code that you are incorporating into your project to look for names that could be made more understandable.
- When possible, change names to match the comments until the comments are no longer necessary.

8 Hardcode

DESCRIPTION

Hardcode refers to placing values that affect the application directly in the code when they would be easier to maintain in separate resource files. While very few would consider placing the data for an image directly in the source code, it is common for a string or number to be placed directly in the code.

Hardcoding values into the source code of an application can cause many problems and solves few. Yet, this quick and lazy solution is taken far too often as the path of least resistance. The proper handling of these values represents hard work, and it is therefore not surprising that it is not done properly. However, this hard work will pay for itself and more as development moves along.

SYMPTOMS

Hardcoded values are in general easy to spot in the source code itself. Nevertheless, to further the motivation to remove them, let us look at some of the results that occur from hardcoding these values.

Strings and Numbers

The two most common value types that are hardcoded are strings and numbers. Very few other base types are hardcoded, and many of the others that are represent a collection of strings and numbers. With only a few exceptions that we will mention shortly, any string or number that is placed directly in the code is likely to cause the problems indicative of the Hardcode illness.

What are the exceptions to the rule of no strings or numbers in the code? First, let us consider strings. There is really only one string value that should ever be placed directly in the code, and that is the empty string (`""`). Any other string, from one letter to a full book, should be stored in an external resource file.

The case for numbers is slightly more complicated. Determining exactly when a number can safely be hardcoded is a matter of context. In general, however, only the numbers 0, 1, and 2 should be found directly in the code. Most valid cases for hardcoding these values involve either a comparison, such as determining if a number is zero or not, or as part of a fixed equation, such as inverting a number. The reason that these hardcoded values are acceptable lies in the fact that the equation or comparison is extremely unlikely to change later in development. Therefore, any number that is likely to change, even if it is 0, 1, or 2, should not be hardcoded.

The following lines of code illustrate both acceptable and unacceptable use of numbers in code:

```
if(value > 0) {                          // acceptable
float inverse = 1.0f / n;                // acceptable
currentSpeed = 1.0;                      // unacceptable
Vector midpoint = (a + b) / 2.0;         // acceptable
double area = 3.1416 * r * r;            // unacceptable
```

Notice that the use of `1.0` as a speed is not considered acceptable because it is not part of a fixed formula. The `1.0` could be replaced with any number and should therefore not be hardcoded. On the other hand, both the `2.0` and `3.1416` are part of a formula. However, since `3.1416` represents a mathematical concept, it would be better to give it a name such as `PI`. The `2.0` is not as easily named and is simple enough that it can be allowed to stay hardcoded.

More CAP

One of the most dangerous results of hardcoding is a major illness known as the Cut-and-Paste (CAP) Epidemic. Because the values are not stored in a variable, any usage of the same value must be cut-and-pasted. Thereafter, any change to one of these values must be followed by a change to all of them or unpredictable behavior will result. This is true of numbers and strings, as well as any other odd type that might occasionally be hardcoded.

One solution that is better, but still not optimal, is to place these values in a variable whose scope is accessible to any that need its value. This will eliminate the cut-and-paste that is required of hardcoded values, but it will not solve some of the other problems caused by hardcoding. This solution is most acceptable when the value is extremely unlikely to change, such as the value for p. The other common acceptable usage comes at the end of development, when it is clear that placing the value within the code will allow the compiler to optimize the application in ways that would otherwise not be possible. Be careful to only use this solution when the profiling shows that access to a particular variable is a major performance cost.

Localization

Very little software development these days escapes the need to be international, and nothing can make this more difficult than hard-coded strings. While there are other considerations to making an application suitable for a particular locale, the most prominent is always text. All visible strings must be translated into the various languages of each locale to which the application will be distributed. Even strings used for configuration, which are visible only to advanced users, might require translation.

Strings that must be translated to other languages offer several special challenges to the application's programmers. First, translation is not often done by the programmers, and therefore not only should changes not require recompiling, but the strings should be collected into one convenient location. This makes it obvious which strings must be translated and saves time and money on the translators and on the quality assurance that must check the results.

The other challenge to internationalization of strings is the differences in grammar across languages. Therefore, it is not sufficient to avoid hardcoding the strings. It is also important to be careful about the methods used to construct strings that are generated at run time.

Take, for example, one method of constructing strings that is commonly used in the Java language, which is concatenation:

```
String sentence = "Do any languages treat the " +
    "important " + "arrangement " +
    "of words differently?"
```

Or for the more performance conscious:

```
StringBuffer string = new StringBuffer();
string.append("Do any languages treat the ");
string.append("important ");
string.append("arrangement ");
string.append("of words differently?");
String sentence = string.toString();
```

Do not be concerned, this is not a case of Premature Optimization if you use the second method since it does not sacrifice readability or robustness to any large degree. However, this method of creating strings creates problems when constructing strings that must be localized later. For instance, if you have a lookup function with the following signature:

```
public class Localize {
    static String lookup(String label);
}
```

Then you might be tempted to use the following code to construct a sentence that varies at run time:

```
StringBuffer string = new StringBuffer();
string.append(Localize.lookup("SENTENCE_BEGIN"));
string.append(Localize.lookup("MODIFIER"));
string.append(Localize.lookup("WORD"));
string.append(Localize.lookup("SENTENCE_END"));
String sentence = string.toString();
```

The problem arises when you attempt to translate the English version to French. To understand this, here is an example of the English language lookup table:

```
SENTENCE_BEGIN = "Do any languages treat the "
MODIFIER       = "important "
WORD           = "arrangement "
SENTENCE_END   = "of words differently?"
```

This creates the following sentence:

```
Do any languages treat the important
arrangement of words differently?
```

If you translate the lookup table to French, you get:

```
SENTENCE_BEGIN = "Est-ce que des langues traitent "
MODIFIER       = "important "
WORD           = "l'arrangement "
SENTENCE_END   = "des mots différemment?"
```

Unfortunately, the resulting string is:

```
Est-ce que des langues traitent
important l'arrangement des mots
différemment?
```

Instead, it should be:

```
Est-ce que des langues traitent
l'arrangement important des mots
différemment?
```

Therefore, seeing string concatenation is often a sign of hardcoding. Even though the actual strings have not been hardcoded, the format of the string has been hardcoded. This example does not even mention the further complications that can arise from context-sensitive translation. For many applications, the use of a format string with replacement will suffice, but even that can be considered hardcoded if the string creation has words whose translation are determined by other contextual word replacements elsewhere in the string. Therefore, you must not only keep an eye out for hardcoded strings, but also the method of constructing strings.

PREVENTION

Hardcoding often appears to be the easier and faster method to implementing code that requires numbers or strings. The first step to preventing hardcoding is to realize that this is an illusion, and more often than not it will cost you time down the road. Nonetheless, this might not be enough to overcome the temptation if there is not a convenient method for moving the value to an external file. Therefore, to further decrease the likelihood of hardcoding, we must examine methods to assist in creating and managing external sources for the values that would otherwise be hardcoded.

Standard Data Source

The first step to assisting in removing hardcoded values is to establish a standard location for each type of data. For example, strings that are to be localized are stored in a single file or a group of files separated under a single directory. The file or files generally contain matching label and string pairs. When localization is required, the translator only needs to go to the file or files and change the string values in each pair. This provides a clear advantage to the translator, but does it assist the programmer? The programmer also now has a central location where he can add new strings as needed, and at the end of the project the programmer will not have to go scrounging around to find all the strings that need to be translated. Hence, both the translator and the programmer save time in the long term, even if there is an extra bit of work in the short term.

Not surprisingly, most strings that are to be localized will appear in the user interface. This is true of a large number of values that would otherwise be hardcoded. Placing user interface values outside of the code allows the visual appearance of the application to be tweaked without recompiling. This becomes especially important on large projects with slow build times. It is common to find tools for creating user interface layouts that allow easier manipulation of these external values, but the values are still likely to be external. If you do encounter a user interface builder that places the values directly in the code, be careful, as this is not quite as clean and can slow the testing and refinement process. Occasionally you will encounter user interface components that can be edited while the application is running. This is the best scenario to have during development as it greatly increases the speed of value tweaking, which can be a tedious and time-consuming operation even in the best circumstances.

Both the user interface values and other global application settings are best stored in external configuration files that are, at least during development, text based. This allows easy debugging and sanity checking when necessary. While you can safely convert these files to binary at the end of development, it is an unsafe approach to use the binary files during development except when necessary. Using binary files causes reliance on the tools and programs that edit and convert the binary files. This becomes another failure point that is harder to debug without easy-to-read asset files.

Notice that we have talked primarily about configuration values. These values are susceptible to hardcoding, and this is what we want to avoid. Other forms of data that the application is created to manipulate do not fall under this temptation because that data does not exist as a single value. It is unreasonable to hardcode values when you cannot predict what they will be, thus making these values safe from the dangers of this illness.

XML

Modification and debugging of external values can be made easier by maintaining only a small set of file formats. While you might not be in control of all the file formats in use, due to the tools and third-party libraries used, you should strive to maintain consistency across the file formats that are under your control. This means less syntax to learn for all the programmers on the team, and fewer parsers to debug when things go wrong.

We have already mentioned the benefits of text, because it can be read by a human for debugging what the application and data editors are doing. There is a further advantage, which is the availability of already written text parsers. This can be taken one step further by using XML, which provides a standard data format that is being widely adopted across the computer industry. By using XML, you immediately add a large number of tools and software technologies that can be used to assist in development.

XML is used extensively by many new technologies such as .NET and the Apache Ant build tool for Java. The build file for all the Java examples in Source/JavaExamples/build.xml on the companion CD-ROM provides an excellent example of XML in practical use:

```xml
<?xml version="1.0"?>

<!-- Build file for examples from
     Preventative Programming Techniques -->
     <project name="Prevention"
         basedir="." default="all">

     <!-- Import system environment variables. -->
     <property environment="env"/>

     <!-- Define common properties used
          throughout build file. -->
     <property name="jarfile"
         value="PreventionEx.jar"/>
     <property name="manifest"
         value="MANIFEST.mf"/>

     <!-- Compile the examples using
          the AspectJ compiler. -->
     <target name="compile">
         <taskdef name="ajc"
             classname="
             org.aspectj.tools.ant.taskdefs.Ajc
             "/>
         <ajc source="1.4" srcdir="."
             destdir="." debug="true"
```

```xml
                    deprecation="true"/>
    </target>

    <!-- Package the compiled examples
         in a compressed jar file. -->
    <target name="jar" depends="compile">
        <jar jarfile="${jarfile}"
            manifest="${manifest}"
            compress="true"
            basedir=".">
            <exclude name="build.xml"/>
            <exclude name="**/.nbattrs"/>
            <exclude name="**/*.java"/>
            <exclude name="**/*.form"/>
            <exclude name="${manifest}"/>
            <exclude name="${jarfile}"/>
            <exclude name="apidoc"/>
        </jar>
    </target>

    <!-- Perform the most common
             build functions. -->
        <target name="all" depends="jar"
            description="Build everything.">
        </target>

        <!-- Run a test of the examples
            after compiling them. -->
        <target name="test" depends="all"
            description="Test examples.">
            <java
                classname=
                "com.crm.ppt.examples.Main"
                fork="true" failonerror="true">
                <classpath>
                    <pathelement location="."/>
                    <pathelement
                        location=
                "${env.AJHOME}\lib\aspectjrt.jar"
                        />
                </classpath>
            </java>
        </target>
```

```
<!-- Generate documentation
    from comments. -->
<target name="javadoc"
    description="Javadoc for Prevention.">
    <mkdir dir="apidoc"/>
    <taskdef name="ajdoc"
        classname="
        org.aspectj.tools.ant.taskdefs.Ajdoc
        "/>
    <ajdoc
        packagenames="com.crm.ppt.examples.*"
        destdir="apidoc">
        <sourcepath>
            <pathelement location="."/>
        </sourcepath>
    </ajdoc>
</target>

<!-- Delete generated files. -->
<target name="clean"
    description="Clean all build products.">
    <delete>
        <fileset dir=".">
            <include name="**/*.class"/>
        </fileset>
    </delete>
    <delete file="${jarfile}"/>
    <delete dir="apidoc"/>
</target>

</project>
```

Further examples of the extensive use of XML can be found from Integrated Development Environment (IDE) products such as IntelliJ's IDEA to asset files for games. In particular, XML lends itself to storing any type of hierarchical data that can easily be represented in text format. It is also designed with the ability to read and process the XML file from a stream if the data is structured appropriately, which is important for many applications including Web-based applications.

In addition to the many editors, such as XMLSpy (Figure 8.1), that are available to edit XML, other technologies improve the usefulness of XML. One important addition is XML Schema that defines rules for the structure of a particular data file.

When moving from hardcoded values that can only be changed by programmers to external files editable by other developers, the loss of control can make it easier for errors to be introduced into the data. By providing appropriate XML Schema,

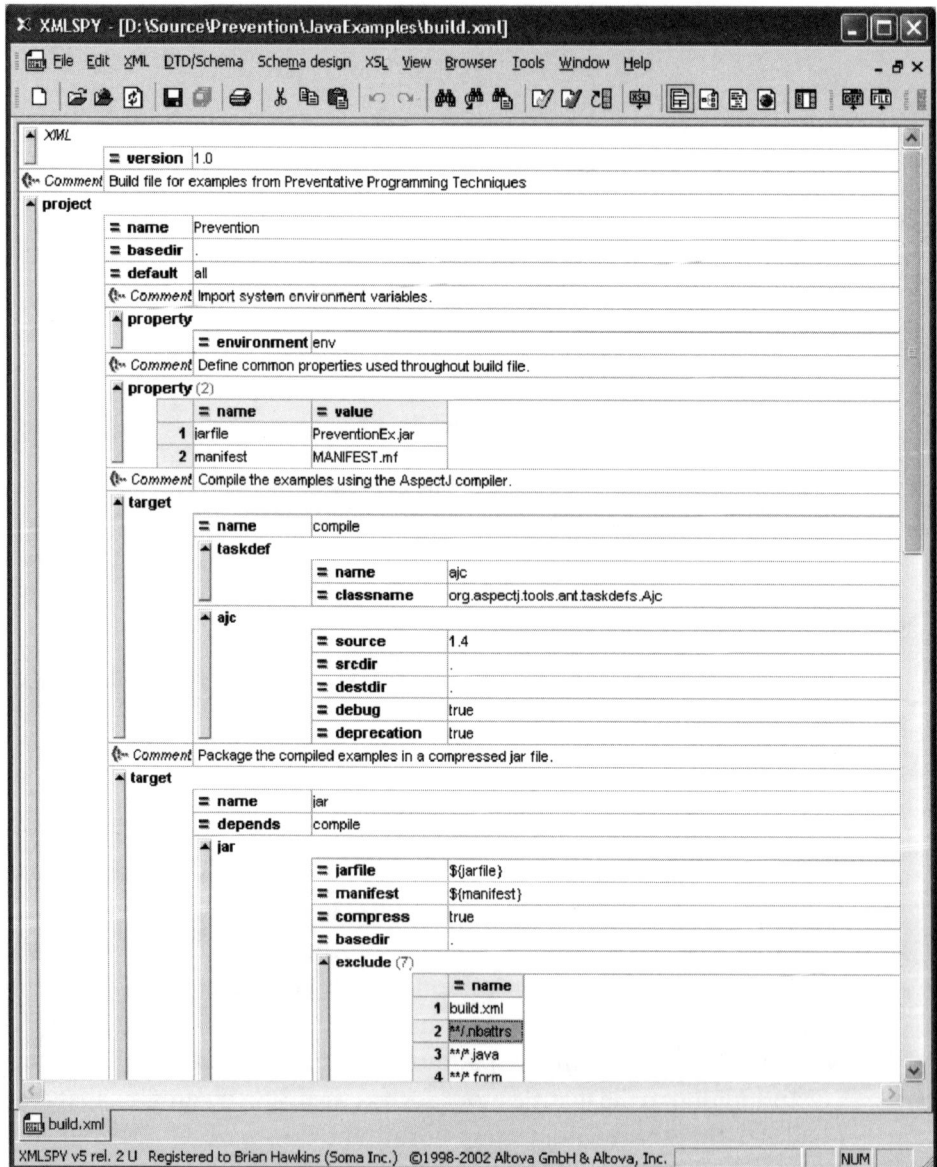

FIGURE 8.1 Ant build file shown in XMLSpy for viewing and editing.

errors can be tested for while editing the file without the need to build complex validation code into your application. It is interesting to note that XML Schema are stored in the XML format as well, allowing them to be edited in the same manner as the other XML files that they will be checking.

The extensibility of XML and the wide array of tools available make it an excellent choice for a wide range of data storage, but be careful not to become enthralled with the virtues and ignore simpler solutions when appropriate. Sometimes all that is needed is a configuration file containing name and value pairs, and XML only reduces the readability. Evaluate your needs and then you can conclude whether XML is the right choice for the particular job you have in mind.

Asset Management

Another important consideration in easing the process of avoiding hardcoded values is management of the resulting assets. Updating and accessing these assets must be seamless if a programmer is expected to add values to them on a regular basis. Any hindrance will increase the likelihood that temptation will win over and the easy route of hardcoding will be taken.

Of primary importance is allowing the simultaneous access and modification of the asset files by multiple programmers. This requires two major components, the first of which is a source control system that supports merging of files edited by more than one programmer at the same time. This presents another important reason for using text files as the basis for these assets, because merging binary files is not a feature that you can expect your source control system to support. Even after a source control system is in place that supports merging, the file format must be such that merging is as unambiguous as possible. Distinct tags can help with this process by preventing matches from occurring by mistake, and XML represents an excellent choice for providing these tags.

Another consideration is the addition and removal of information from the asset files. This must often be synchronized with the changes to source code. Some effort should be made to maintain backward compatibility and version information, but this must be balanced with the need to move forward. Without the backward compatibility testing, failures that occur outside of a single build become problematic. Whether this is possible and whether it is worth the effort must be considered on a case-by-case basis. When backward compatibility is not possible, it is beneficial to have a source control system that supports the submission of multiple files at once to maintain a consistently working build. This is also a reason to avoid partial code submissions to the source control database when possible. It is easy to overlook dependencies between files that are submitted and those that are not. When you must submit only part of the code and assets you are working on, be careful to review what changes were made and look for any missed connections.

Finally, it is beneficial to have an organization to the assets that is easy to understand and extend. This includes directory hierarchy and file naming conventions. Add this to the other asset management concerns not related to the hardcoding and you will often find that asset management can be a full-time job. The amount of time required is an important consideration in scheduling and resource management that should not be overlooked.

CURE

How often have you been asked to work with code that contains hardcoded values? Perhaps even you have written some of these hardcoded values. As these values sit there, particularly those that are duplicated multiple times, they represent errors waiting to happen. These values must be removed from the code before they can cause any problems, or possibly any more problems.

When to Refactor

The simple answer is that all hardcoded values should be removed from the code, but this can be naïve in the face of schedule pressure. Therefore, it is beneficial to consider in what order to tackle this removal. This will leave you with the least risky hardcoded values if schedule pressure forces you to leave some of the hardcoded values in the code.

The highest priority should be given to any strings that must be localized, unless you are working under some strange project that has no possibility of being internationalized. Even if the application is not initially targeted for the international market, it is a good idea to prepare for this eventuality. Otherwise, it can be a tedious process later when all the strings must be extracted from various assets and code to a central file for localization.

Of equal importance are any values that might change when ported to other platforms or languages. These are mostly made up of user interface values, and are less likely than strings to be hardcoded. It is nonetheless a wise idea to ensure that they are not hardcoded and are easily editable to produce different builds for the application's target platforms or possible future platforms.

Only slightly less important are any values that are likely to be tweaked repeatedly. Once again, many of these are likely to be user interface values, but there might be other values as well. By removing these values from the code, the tweaking process can occur at a much faster rate and does not require the assistance of a programmer. This will often save both time and money for the project.

Context Highlighting

Before the hardcoded values can be removed, they must be found. Utilities for finding the values are not commonly available today, but there are still tools that can help with this process. One such utility is built in to many modern code editors, which allow you to highlight the source code text according to the context of each value. In other words, different constructs of the programming language can be rendered in different colors and fonts to make them more distinguishable. For our purposes, the most interesting of these are the highlighting of numbers and strings (Figure 8.2, page 246). By choosing a bright noticeable color and bold font, these values can be made to stand out against the surrounding code. This is an excellent way, while browsing through the code, to spot hardcoded values that need refactoring.

Context highlighting can also be applied to code that is printed out. This provides a much better medium if there is a large amount of code to browse, but color printing is still not common so font highlighting must be used in many cases (Figure 8.3, page 247). This process can still be somewhat time consuming, so if you are pressed for time but still need to remove hardcoded values, then context highlighting will be of less help. One editor tool that can still help is regular expression searches. With a little bit of work, a regular expression that finds number and strings while ignoring other language constructs can be constructed. This is particularly useful if the search can be applied across multiple files. This limits the amount of code that must be visited, and combined with context highlighting can provide a fast and effective way to catch most hardcoded values.

Collection

Removing the hardcoded values from the code is best done as a two-step process. First, collect each value into a global variable or other variable with a wide scope. Many languages offer a method of marking this variable as unchanging, such as

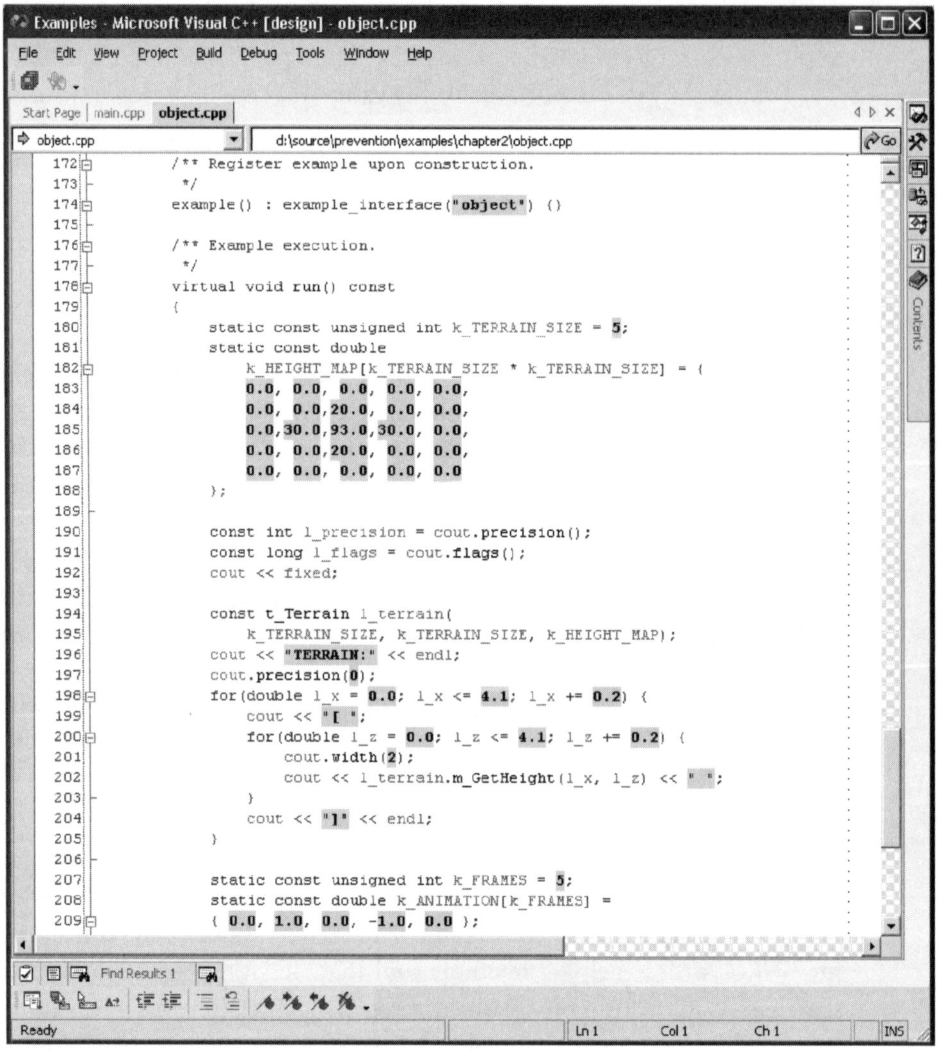

```
 Examples - Microsoft Visual C++ [design] - object.cpp          _ □ X

File   Edit   View   Project   Build   Debug   Tools   Window   Help

Start Page | main.cpp | object.cpp                              ◁ ▷ ✕

⇨ object.cpp              ▼    d:\source\prevention\examples\chapter2\object.cpp      Go

 172   /** Register example upon construction.
 173    */
 174   example() : example_interface("object") ()
 175
 176   /** Example execution.
 177    */
 178   virtual void run() const
 179   {
 180       static const unsigned int k_TERRAIN_SIZE = 5;
 181       static const double
 182           k_HEIGHT_MAP[k_TERRAIN_SIZE * k_TERRAIN_SIZE] = (
 183           0.0, 0.0, 0.0, 0.0, 0.0,
 184           0.0, 0.0,20.0, 0.0, 0.0,
 185           0.0,30.0,93.0,30.0, 0.0,
 186           0.0, 0.0,20.0, 0.0, 0.0,
 187           0.0, 0.0, 0.0, 0.0, 0.0
 188       );
 189
 190       const int l_precision = cout.precision();
 191       const long l_flags = cout.flags();
 192       cout << fixed;
 193
 194       const t_Terrain l_terrain(
 195           k_TERRAIN_SIZE, k_TERRAIN_SIZE, k_HEIGHT_MAP);
 196       cout << "TERRAIN:" << endl;
 197       cout.precision(0);
 198       for(double l_x = 0.0; l_x <= 4.1; l_x += 0.2) {
 199           cout << "[ ";
 200           for(double l_z = 0.0; l_z <= 4.1; l_z += 0.2) {
 201               cout.width(2);
 202               cout << l_terrain.m_GetHeight(l_x, l_z) << " ";
 203           }
 204           cout << "]" << endl;
 205       }
 206
 207       static const unsigned int k_FRAMES = 5;
 208       static const double k_ANIMATION[k_FRAMES] =
 209       { 0.0, 1.0, 0.0, -1.0, 0.0 );

 ☑ 目 🔄  Find Results 1  🔄
 🖳 🖳 🖳 A⁺ 🖳 🖳 🖳 🖳 ◢ 🖳 🖳 🖳 .
Ready                                    Ln 1      Col 1     Ch 1        INS
```

FIGURE 8.2 Context highlighting is used to make strings and numbers more visible in Microsoft Visual Studio.

const in C++ and final in Java, and always available, such as static in both C++ and Java. These modifiers can be applied for both readability and performance reasons.

Preferably, one central location can be used initially or at least only a few collection locations. As similar values appear in the collection, they can be examined

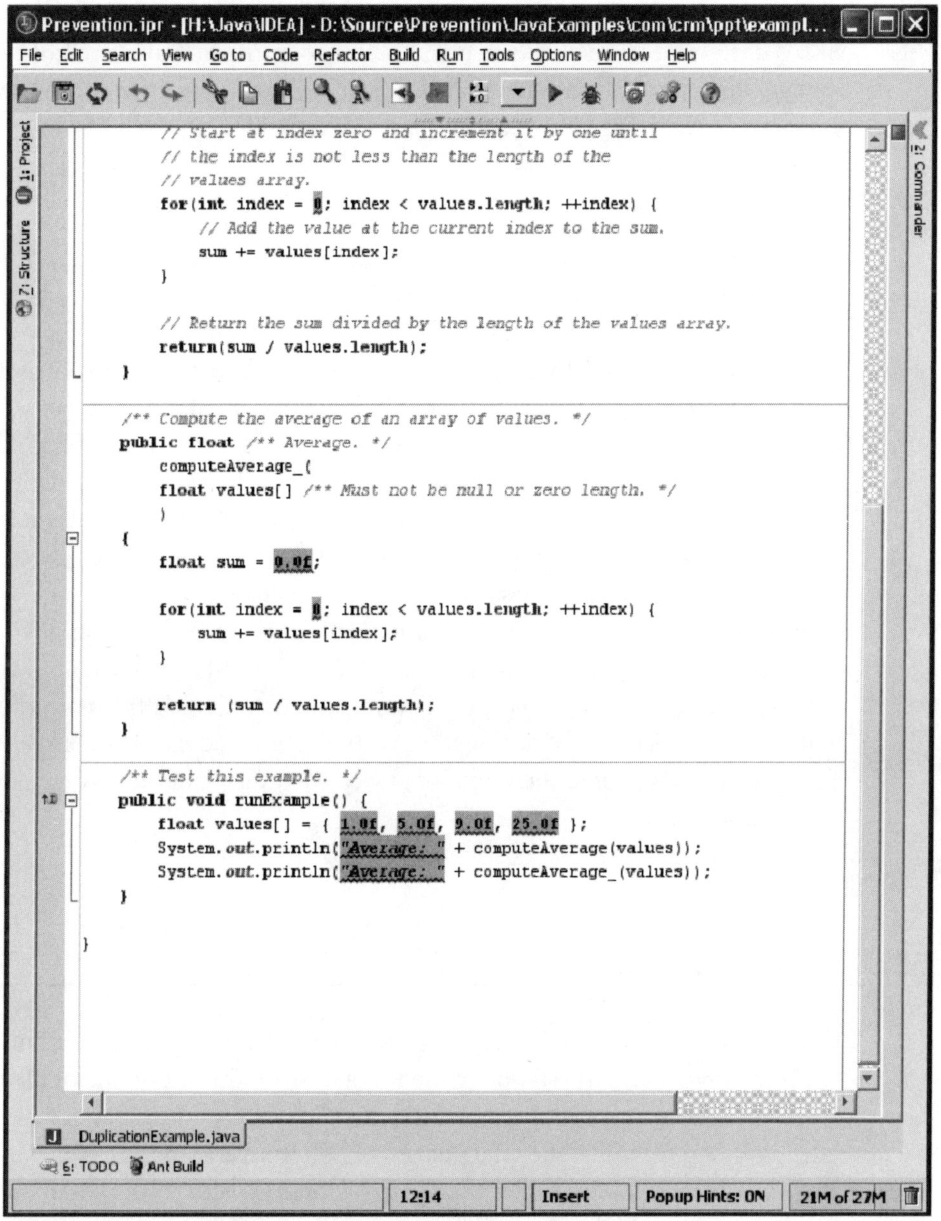

FIGURE 8.3 By adding underlining to strings and numbers in IntelliJ's IDEA, they become more visible even when printed.

to determine if they represent the same concept. Do not be tempted to combine values simply because they are equal. Later, one of the values might require a change while the other should remain the same. Combining different constants that happen to have the same value will make this difficult.

Run Time vs. Compile Time

Once all the hardcoded values have been collected and collated, a decision can be made about their final resting place. The first major decision that must be made for each value is whether the value should be in an asset file or remain in the code. Most values should be moved to an asset file, but there might be occasions when the value cannot change except under the most unlikely circumstances. For example, the value for π is extremely unlikely to change. Because the extra work will provide little chance of future gain, it is reasonable to leave these values in the code. Nonetheless, they should still be placed in meaningful locations and retain a readable name, thus making the code that uses the value more readable.

The majority of the values can then be moved to an asset file. These values are then read in, often at initialization time, and can be placed in a variable with the same name that was used to collect the value. This limits the amount of code that must be changed in each step of the process, thus reducing the amount of error that can occur. As always, testing after each change is advisable to ensure that the changes have not altered the values of any calculations, and therefore the behavior of the application.

RELATED ILLNESSES

One of the common excuses used to motivate hardcoding is optimization, but this really represents a manifestation of the Premature Optimization major illness. Even when optimization is a concern, which it should be only at the end of development after profiling carefully, most languages still offer several mechanisms for naming and sharing these values rather than placing them directly at the end location. Regardless, this is only the case in a few select instances. The rest of the time, concern should be given to development speed over optimization.

At this point, some programmers might claim that organizing and sharing the values takes time and therefore costs development speed. This is an unfortunate symptom of Myopia, which causes the programmer to only think in the short term.

These hardcoded values will cost on average more time later in development; therefore, the long-term view supports removing the hardcoded values.

FIRST AID KIT

With XML becoming an industry standard for many types of data files, it is important to consider its use on your project. One advantage that XML has is the variety of tools available for reading, editing, and manipulating it. Most languages have freely available libraries that implement several varieties of XML parsing along with other utilities such as XML transformations using XSLT. In addition, products such as XMLSpy provide advanced tools for viewing and editing any XML file in addition to extra support for standard formats built on top of XML.

Another useful tool for spotting unwanted strings and numbers in code is context highlighting for editors. Most major editors, including Emacs, Microsoft Visual Studio, and IntelliJ's IDEA, support context highlighting and allow different fonts and colors for strings and numbers to make them stand out.

SUMMARY

The symptoms of Hardcode are:

- Numbers and strings that are embedded directly in the code.
- Numbers and strings that are cut-and-pasted to multiple locations in the code.
- Localization requires changes in the code and in multiple locations.

To prevent Hardcode:

- Standardize location and access methods for numbers, strings, and any other configurable data.
- Prefer XML if it is appropriate because of the wide range of tools available.
- Standardize asset creation, locations, and management.

To cure the effects of Hardcode:

- Refactor localized strings and porting values first.
- Refactor commonly changed values to a data source outside the code.
- Use context highlighting available on most modern editors to assist in spotting hardcoded string and numbers.
- Collect any repeated values in a single location in the code.
- Move most of these values to a source outside the code, leaving only those determined to be performance critical in the code.

9 Brittle Bones

DESCRIPTION

Just as a strong foundation is required to erect a solid building, so is a strong foundation necessary for creating quality software. Without such a foundation, a building might collapse at any point, and the same is true of software development. A single error in the initial code can cause a cascading effect as more and more code is added on top of it.

Unfortunately, many factors conspire to cause a programmer to continue ahead on top of a weak foundation. Poor functionality or a total lack of certain functionality is unwisely seen as a problem that can be remedied later. As the program grows in a more organic fashion, control is lost and becomes difficult to recover.

Take, for instance, one project that was displaying a map of tiles from an isometric viewpoint. While building the initial foundation for the application, a decision was made to represent the tiles using a diagonal layout in memory thought to be similar to the layout of the display. This small decision would lead to the loss of countless hours of development time.

The first signs of trouble appeared when it became necessary to measure the distances of paths traced through connected tiles. Because tiles that were adjacent in the display were not necessarily adjacent in memory, the indices could not be used in a straightforward manner to determine the distance between tiles. This was further complicated by the difficulty in even following a straight-line path through the data in memory. This led to a different programmer implementing different methods for handling this bizarre scheme. The ensuing plethora of approaches was hard to maintain and errors fixed in one were not fixed in another.

One such approach was particularly problematic because it duplicated the data in a separate memory structure with a more straightforward layout. While this simplified algorithm used this separate data set, it was difficult to keep the two sets of data properly synchronized. It also represented an unnecessary waste of resources without any of the gains that could justify having two sets of data. This one single decision was responsible for hours of lost time, and consequently lost profit.

SYMPTOMS

The symptoms of Brittle Bones are much subtler than many of the other illnesses. Often times, they will not reveal themselves until later in development, when a solution is much more difficult to find. Therefore, it is important to look carefully for small signs as they appear. This will allow the problem to be remedied before it becomes a monumental task.

Too Minimal

One sign of a weak foundation is the repeated implementation of similar functionality on top of that foundation. This is similar to erecting a building in a swamp and then piling on extra layers of material to try to fix the sinking. Eventually the house might stabilize and the upper floors will be dry, but the foundation required considerable money and material, which only resulted in a watery foundation. A similar phenomenon occurs with code; as functionality is piled onto the weak foundation, it becomes susceptible to error or complete failure. Even if the resulting application eventually works, programmer time and money were wasted on unnecessary extra work.

Just such a poor foundation class was found in one project that decided against the use of the C++ Standard Template Library. This was already a sign of NIH Syndrome, but the situation was worsened by the implementation of a homegrown list class that was far from complete. The list class that was created implemented only the most basic features such as adding and removing an item from the list. A basic list iterator was also created to go through the list, but offered little functionality that a simple index would not have provided.

The list class was similar to the following code, which can also be found on the companion CD-ROM in Source/Examples/Chapter9/minimal.h:

```
/**     @warning DO NOT USE THIS CLASS FOR
 *      ANYTHING EVER!!!
```

```
 */
class t_List
{
public:

    class t_Iterator
    {
    public:

        t_Iterator(t_List *list)
        {
            m_list = list;
            m_index = 0;
        }

        void mReset()
        {
                    m_index = 0;
        }

        void operator++()
        {
            m_index = mEnd() ?
                m_index : ++m_index;
        }

        void operator++(int)
        {
            m_index = mEnd() ?
                m_index : ++m_index;
        }

        void operator--()
        {
            m_index = m_index ?
                --m_index : 0;
        }

        void operator--(int)
        {
            m_index = m_index ?
                --m_index : 0;
        }
```

```
void *mValue()
{
    return(mEnd() ?
        0 : m_list->mGet(m_index));
}

bool mBegin()
{ return(m_index == 0); }

bool mEnd()
{
    return(m_index >=
        m_list->mGetCount());
}

int mGetIndex() { return(m_index); }

private:

    t_List *m_list;

    int m_index;

};

t_List(int size)
{
    m_size = size;
    m_count = 0;
    m_data =
        (void**)malloc(m_size *
        sizeof(void*));
}

void mAdd(int index, void *data)
{
    if(m_count == m_size) {
        return;
    }

    if(index < 0 || index > m_count) {
        index = m_count;
```

```
        }

    if(index != m_count) {
        for(int i = m_count; i != index; i--) {
            m_data[i] = m_data[i - 1];
        }
    }

    m_data[index] = data;
    m_count++;
}

void mRemove(t_Iterator *iter)
{
    for(int i = iter->mGetIndex();
        i < m_count - 1; i++) {
        m_data[i] = m_data[i + 1];
    }
    if(m_count) {
        m_count--;
    }
}

int mGetSize() { return(m_size); }

int mGetCount() { return(m_count); }

void *mGet(int index)
{ return(m_data[index]); }

private:

    int m_size;

    int m_count;

    void **m_data;

};
```

The lack of functionality caused the project's programmers to implement their own search routines in different areas of the code, each with its own unique bugs that had to be fixed individually. If instead the functionality had been contained in the list class, only one place would have to be debugged for the entire application to benefit. This extra debugging time was added on top of the wasted time implementing the same solution multiple times, because it was not obviously available. This again shows a combination of illnesses that began with NIH Syndrome and Brittle Bones, and were subsequently complicated by CAP Epidemic and Docuphobia.

Another missing element to the completeness of this list class is the use of templates. Instead, the user of the list is forced to use `void*`. This leads to a multitude of unsafe type casts, and the ensuing errors that these casts always cause. This is particularly problematic with a container class as the insertions can be well separated from the removals that increase the dangers created by the lack of safe typing.

Making this class complete as well as minimal would require substantial additions and changes to the interface. In this case, we are in luck because a minimal and complete interface and the accompanying implementation are available from the C++ Standard Template Library. Although it is not perfect (nothing is) the C++ Standard Template Library provides many examples of well thought out interfaces that fulfill the requirement of being minimal and complete. If you are using C++, this is an excellent set of examples to use in understanding how to make your interfaces minimal and complete. Other languages often have similar libraries that represent good examples of minimal and complete interfaces that can be used for learning to write your own interfaces.

Too Complete

On the opposite side of the coin, it is also possible to provide a foundation that has plenty of functionality stuffed into it, but is still weak and difficult to build upon. One of the biggest problems with too much complicated functionality is the readability of the resulting interfaces and code. As it becomes more difficult to find and understand the functionality available, the programmer must waste time and resources hunting and experimenting to accomplish his goal. Even worse, you might end up with a plethora of methods for accomplishing a task and then find every programmer doing it a different way. This will lead to confusion and more readability issues. Worse still, the programmer might give up altogether and implement the functionality himself. In this scenario, you not only lose the time spent searching and the time spent implementing, you also introduce duplicate functionality that leads to many of the ill effects of the CAP Epidemic.

An example of this type of interface can coincidentally be drawn from a technology that has been redone many times over. This particular complicated interface was part of a user interface library that was being developed to work with DirectX, which did not possess its own user interface functionality. In order to overcome this deficiency, a number of user interface components were created from scratch. One such component was the commonly needed scrollbar. This component is present in almost every windowing user interface and is extremely useful for presenting large quantities of information.

This particular scrollbar was created with, among other things, a horizontal and vertical size. This resulted in a constructor similar to the following:

```
Scrollbar(..., int hsize, int vsize, ...);
```

At first, this might seem like it is necessary to create both a minimal and complete interface, but on closer examination, a problem arises. First, we can create a vertical scrollbar as follows:

```
new Scrollbar(..., 0, 10, ...);
```

In addition, we can create a horizontal scrollbar like this:

```
new Scrollbar(..., 10, 0, ...);
```

However, what happens when we create a scrollbar like this:

```
new Scrollbar(..., 10, 10, ...);
```

This does not make sense using the traditional definition of a scrollbar. Nevertheless, since it was possible to create a scrollbar this way, the library implementer was forced to provide a working control. This control consisted of a large rectangle containing a smaller rectangle that could be moved both horizontally and vertically. While this was an interesting trinket, it was hard to place and of little use in practice. Therefore, although at first glance this interface appeared necessary for a complete interface, the practical uses of scrollbars dictated that this interface violated the principle of a minimal interface and made it too complete.

This interface led to several problems, not the least of which was the loss of time the library implementer incurred by implementing an unused feature. Additionally, the interface was hard to understand, and led to several bugs that could have been avoided. The type of scrollbar created was also not immediately evident to other programmers looking at code that created a scrollbar. A better interface would have been:

```
enum t_ScrollbarOrientation {
    k_IS_HORIZONTAL,
    k_IS_VERTICAL,
    };
Scrollbar(..., t_ScrollbarOrientation orientation,
    int size, ...);
```

This eliminates the need to implement additional library functionality and as a bonus makes the creation code easier to read:

```
new Scrollbar(..., k_IS_HORIZONTAL, 10, ...);
```

This example also illustrates how the problem domain has a large impact on exactly what fulfills the definition of a minimal and complete implementation. Software development is not an end unto itself, but a method of solving problems most often relating to a specific problem domain.

Lack of Consistency

Another problem that can occur when building a foundation for an application is a lack of consistency in the interfaces built and used for the project. When we refer to consistency, we are talking about consistency within the application and consistency with external interfaces as well. Particularly important is consistency with the language standards and community at large. Violating this consistency can lead to confusion throughout the course of development, often enough to introduce major errors into the code.

For example, C++ allows programmers to overload the meaning of many of the built-in operators in order to allow their use with other object types. When used to create syntactic representations that make sense, such as vector arithmetic, this can make the code much more readable. However, you might see the symptoms of Brittle Bones when this overloading starts taking the form of typing shortcuts. Using addition to indicate the intersection of two sets would be an example of a confusing use of this overloading. The reader would have no intuitive indication of the meaning of addition when applied to the two set objects.

If a programmer saw:

```
Set c = a + b;
```

He might be unsure if this meant to add the members of the two sets, take the union of the two sets, take the intersection of the two sets, or perhaps something else entirely. While there might be a reasonable chance that the programmer would pick the correct meaning, the ambiguity could be eliminated with only a few more characters:

```
Set c = a.union(b);
```

The meaning of vector addition, on the other hand, is well accepted; therefore, overriding the addition operator for vectors is reasonable:

```
Vector c = a + b;
```

However, overriding the multiplication operator can be problematic, as some programmers might mistake it for a dot product operator. Thus, even in the simple case of vector multiplication, it is better to spell out the operation than overload an ambiguous operator.

Even consistency in naming is important for a solid foundation. In Java, querying the `boolean` state of a particular aspect of an object is usually done with a function that is prefixed with `is`. Violating this convention is another inconsistency that will lead to problems throughout development. Developers will have to search harder to find the function in an unexpected location, or worse, they might go ahead and write their own version of the function.

Returning to the example given earlier when we talked about overly minimal interfaces, there is also a consistency problem with the operation of the list class. Once again, this relates to the problem of attempting to reinvent the technology already existing in the C++ standard template library. In this case, the iterator class violates the convention used for list iterators in the standard template library that says they are invalidated when the element they refer to is removed from the list. This will cause confusion among the other programmers on the project, as the iterator will not behave as they expect if they have used the standard template library. Because it resembles the iterator from the standard template library, this becomes a major problem. Further, once the developers have adapted to using this new style iterator, they will litter the code with uses that will not allow them to change over to the standard template library despite the fact that it would appear that such a change would be simple. This is made more detrimental because the list class is lacking functionality that could have easily been added by performing such a switch.

Spiraling

Whatever has caused the foundation to become brittle, one symptom that will always manifest is a cascade of problems that build upon each other. This spiraling loss of control starts with the foundation and becomes worse as each layer is built on top of the last (Figure 9.1). Unfortunately, this symptom is not obvious until much later in development and is therefore akin to the final stages of a disease. Just as with a disease, this can easily be a harbinger of the project's death unless immediate action is taken. Even with immediate action, an irrecoverable cost has already been paid at this point. Therefore, be diligent and work to avoid ever seeing this symptom except in its first subtle stages.

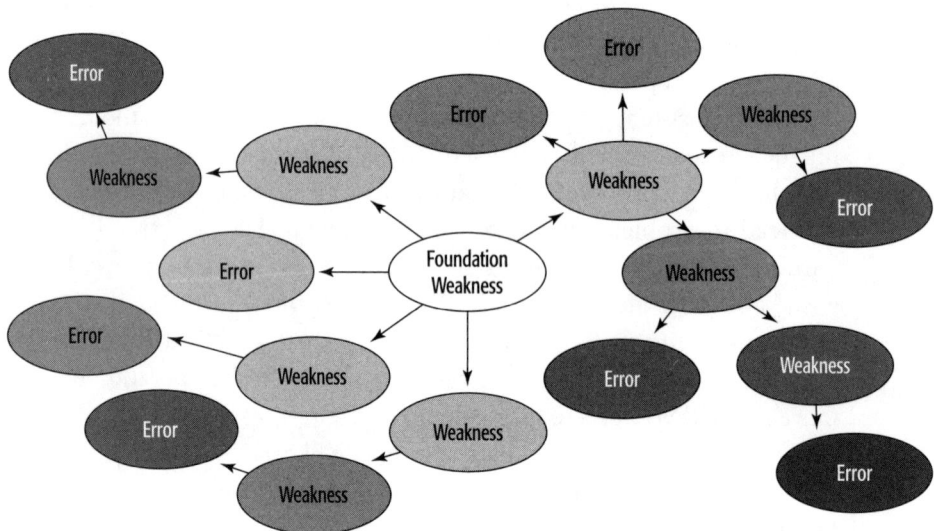

FIGURE 9.1   Illustration of how a single weakness in the foundation can be built upon to create further weaknesses throughout development. This often obfuscates the fact that the weakness originates from the foundation, making this illness even more difficult to diagnose.

PREVENTION

Prevention of this illness is particularly important because of its far-reaching consequences. Although there are steps available to cure Brittle Bones, these solutions incur substantial cost and are not to be relied upon. As a clarification, it is not suggested that no refactoring be done to the foundation. Refactoring is still an important part of the software development process, but for large projects in particular, a certain amount of upfront design is necessary to reduce the risks of everything collapsing later in development.

Design

The amount of design work that should be done on a project is a matter of great controversy these days. With Extreme Programming advocating the new approach of minimal design and the old school engineers advocating a complete design at the start, what conclusion can we come to about how to handle design? The answer is to match the level of design to the project and the team of developers. It is necessary to take into account both the project scope and the skill of each developer.

First, let us discuss how the scope of the project affects the amount of design necessary. For small projects, the number of layers that the programmers on the project are responsible for stacking is also likely to be small. This makes it easy to separate tasks and their underlying implementations such that the initial design can be minimal. As the project gets larger, the number of separate modules and layers increases. This also increases the impact of each mistake made on early code, as it is more likely to affect other code adversely and more likely to be used often. Because of this, a more solid initial design is required (Figure 9.2). This design does not necessarily have to go into details, but should provide enough information to determine what functionality should be concentrated on and what interfaces are required to communicate between modules and layers.

The other primary factor in deciding on the detail of the initial design is the makeup of the team. This is affected by two factors, the size of the team and the skill level of each member. As with the size of the project, the level of detail in the design is directly proportional to the size of the team. This is particularly true of large teams due to the need to break them into smaller teams for better communication. This division into smaller teams can only work if there is a design to guide the integration of code between each of the teams.

As for the importance of the skill level of each team member, the amount of design required is inversely proportional to the skill level of the team members. However, this is not nearly as linear of a function as team size. A team full of disciplined and skilled programmers needs only minimal design, but add one mediocre programmer and the team must adapt with a much clearer design to reduce the need to constantly help the less skilled programmer. Once again, this favors small

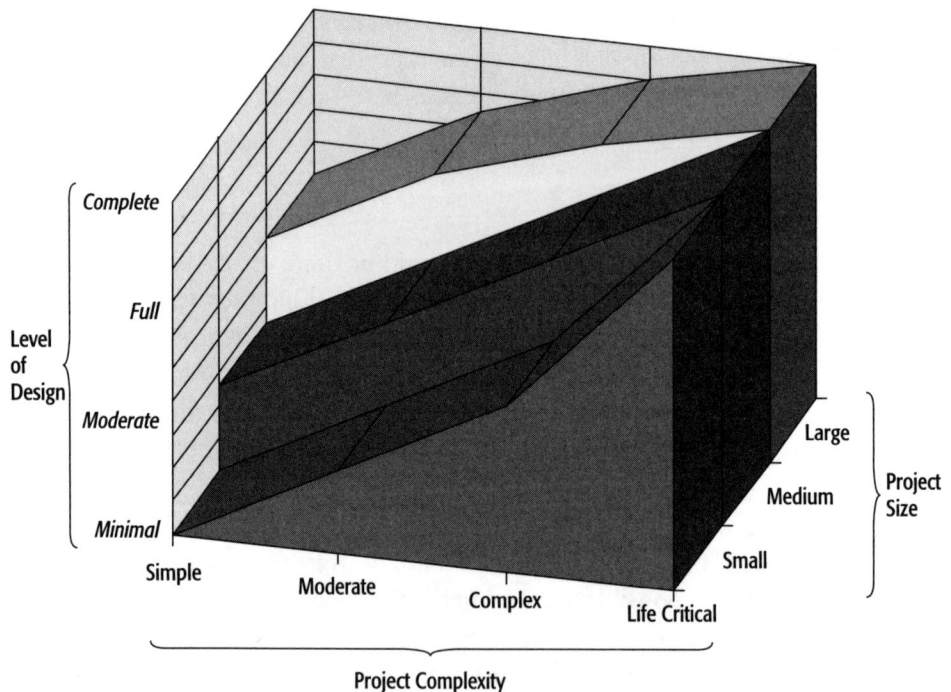

FIGURE 9.2 Effect of project scope and complexity on the amount of initial design required.

projects having less design than large projects because it is much easier to create a small team of skilled developers than to find a large number without introducing some mediocre developers (Figure 9.3).

Design content will vary by project and detail level, but the most important part of the design is always the interaction between code developed by different programmers. This generally takes the form of interface documentation, and while initially it is written as a separate document, it should subsequently be incorporated into the comment documentation as the code is developed. With the design and documentation available early, programmers can spot missing or extraneous functionality and direct their observations to the appropriate team member.

Following these guidelines should help in determining the appropriate design for your project. There are no hard-and-fast answers, but guidelines and experience will help to find the right level of design.

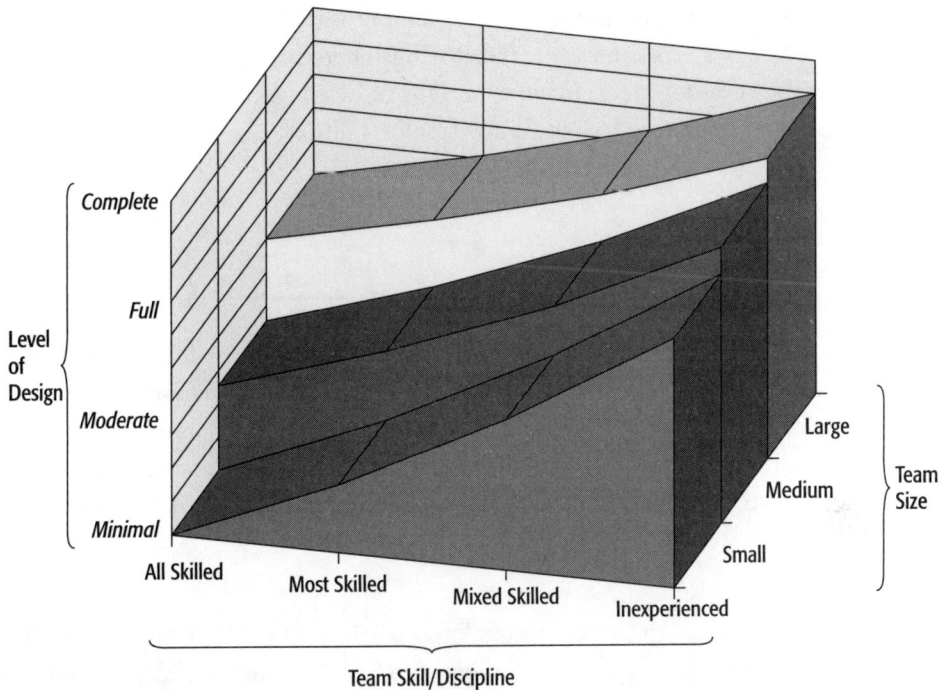

FIGURE 9.3 Effect of team size and skill level on the amount of initial design required.

Balance

The art of developing a good foundation for a project is at the core a balancing act on several fronts. First, you must balance team size versus the skill level of the programmers. Next, you must balance the amount of design work versus the amount of actual code writing. Then, you must balance the amount of functionality versus the complexity of each module. Individual programmers must balance the amount of documentation versus self-documenting code.

As was mentioned in the Introduction, no silver bullet provides the perfect answer to all of these problems. However, that does not mean that there are not better approaches that reduce the risk of making a mistake. Guidelines can help to reduce risk, and that is a primary goal of software development that is too often ignored. Guidelines also give you the confidence to make decisions instead of putting them off because of fear.

Another important point to remember is that *refactoring* is not a bad word. In fact, few projects can escape refactoring. Therefore, do not let the fear of mistakes paralyze development. Instead, aim for reduced risk to eliminate the big problems and use refactoring to handle any smaller failures.

CURE

As was mentioned earlier, the earlier the weakness of the foundation is discovered, the easier it will be to fix. However, not all hope is lost no matter when the problem is discovered. There is always a chance to fix the problem, but be careful not to look for hacked solutions. These will only fix the problem temporarily, and in the end will worsen the situation. Instead, what is called for is a careful and considered approach.

Stop

The very first step to fixing a weak foundation is to stop development. Do not add any more features, and do not attempt to fix bugs unrelated to the foundation. Take a step back from the code and examine the problem or problems that are manifesting repeatedly. Continuing to develop will introduce new code that uses the weakened foundation even as the foundation is changing. This will lead to more errors and lost development time due to the difficulty of coordinating these two separate efforts.

Instead, the team resources should all be directed to fixing the foundation first. If some of the programmers cannot contribute to this process, find work that is not directly related to changing the application. This can also be a good time for them to spend time researching material for use later in development. This will derive at least some benefit from the time lost refactoring the foundation.

Learn

The next step is to determine what sections of the foundation are causing repeated problems. The goal of refactoring the foundation is to prevent more problems from occurring, and this can only be done by analyzing the current problems. It is important during this step to resist the temptation to make changes, as these changes might only fix one of the symptoms and not the disease.

Once all the symptoms that lead to the suspicion of Brittle Bones are collected, foundation changes can be considered to fix both the symptoms and likely future problems. This requires some extrapolation and is one of the reasons for collecting all the problems together. Much can be learned from similarities between symptoms, and this can lead to a shared solution. The more symptoms a single solution cures, the more likely that solution is to prevent future problems.

Refactor

At this point, you should have a list of possible solutions to implement. Now the actual refactoring must be done. Start by prioritizing the solutions in order of their impact; the more a solution is expected to fix the earlier it should be implemented. Now implement each solution one at a time. After every solution is implemented, test that the application functionality has not changed and that the solution fixed the expected symptoms. By keeping a tally of the symptoms that are fixed, you can be sure that all necessary problems are removed once the refactoring is complete.

RELATED ILLNESSES

Brittle Bones is related to all three major illnesses and a couple of minor illnesses as well, thus reflecting its far-reaching nature. All three major illnesses can lead to a brittle framework that is the major symptom of Brittle Bones. First, Premature Optimization causes unnecessary dependencies to be formed between different sections of code. This reduces encapsulation and abstraction, and subsequently weakens the base framework. Cut-and-paste programming also leads to a poor foundation that is difficult to maintain. Functionality is likely to change in one place but not in another, thus causing odd behavior for which it is difficult to find the cause. Cut-and-paste also limits the reusability of code in the foundation, making it less complete. Finally, being struck by NIH Syndrome means that the entire foundation must be written by the team. This increases the time to develop the foundation and the need to test thoroughly or risk errors. Using third-party solutions can greatly accelerate the initial foundation work, and these solutions should be seriously considered.

Brittle Bones can also be part of the cause for a CAP Epidemic. An incomplete foundation will require extra duplication of structures on top to make the code base stable. This duplication will generally result in many of the symptoms that plague cut-and-paste, and will often be the direct result of using the cut-and-paste feature. Keep a careful eye out for this so that it can be cured promptly.

The two minor illnesses that can cause Brittle Bones are closely related themselves. Complexification can cause a foundation that is overly complex and difficult to use, or in other words, a foundation that is too complete. Over Simplification causes a foundation with the opposite problem, a framework that is too minimal. Both of these lead to a cascade of errors due to missing or misunderstood functionality. As with all other causes of Brittle Bones, these must be remedied quickly to prevent further costs.

FIRST AID KIT

The primary tools for avoiding Brittle Bones are good requirements and design tools. Given the wide range of methodologies available for projects, the number of tools for assisting in the requirements and design phases are large and varied. Since the methodology chosen is and should be project specific, the best choice of tools is also methodology specific.

SUMMARY

The symptoms of Brittle Bones are:

- A large number of missing features in code that has or will be used as a basis for other code, making the code too minimal or overly simple.
- An abundance of unused or overly complicated features in code that has or will be used as a basis for other code, making the code overly complete.
- A myriad of different styles and components that are difficult to keep track of and understand.
- A feedback cycle of workarounds and poor code that creates more workarounds and poor code until the system is unmanageable.

To prevent Brittle Bones:

- Perform sufficient initial design based on project type, team makeup, and methodology chosen.
- Look for a balance between design work and coding that provides a solid foundation without excess time wasted.

To cure the effects of Brittle Bones:

- Stop working on new features when a flaw has been found in the foundation code.
- Analyze the code to discover all problems with the foundation code.
- Refactor the foundation code with the new knowledge gained from the problem analysis.

10 Requirement Deficiency

DESCRIPTION

Unstated requirements can cause considerable damage, particularly if they are not caught until later in development. This Requirement Deficiency is a common problem across all types of projects, particularly in certain areas of development. At first it might seem that this is only important to technical leads and other higher up personnel, but on closer inspection individual tasks and development efforts must also fit within the larger requirement structure. Thus, individual programmers can be seen as working within a subset of the requirements, making at least part of any problems caused by missing requirements to affect their work as well.

SYMPTOMS

If detected early in development, dealing with missing requirements is as simple as adding them to the design, and ensuring that they are taken into account at the appropriate locations. This means that detecting the absence of these requirements is particularly important to achieve early in the project. Let us look at symptoms that can indicate Requirement Deficiency.

Vagueness

The most general symptom is vagueness in one or more areas of the current requirements. While it can be used to spot possible missing requirements, this vagueness unfortunately has the opposite effect on many occasions. The small bit of mostly useless information can obfuscate the need for more detailed requirements about a topic.

For example, here are some vague requirements that can easily end up as the final requirements:

- The application must perform at a reasonable level.
- The application must use as little memory as possible.
- The application must be completely bug-free.
- The application must be ready for possible changes.

These requirements can easily make their way into the final requirements for a project, and only later will the important questions arise about what exactly these requirements mean. Not only are these requirements vague, but they leave plenty of room for interpretations that are very unrealistic for a developer to achieve.

If you immediately have questions about a requirement that are not answered anywhere else in the requirements, then it is likely that the requirement is too vague. The examples of vague requirements given earlier would quickly bring to mind several looming questions:

- What is a reasonable level of performance?
- Is there a particular limit to the memory that can be used?
- What does bug-free mean and what can be used as proof that no bugs exist?
- Will the application need to provide and answer to life, the universe, and everything?

Two important notes about this particular symptom deserve attention. First, while vague requirements can point to missing requirements, some missing requirements will be absent in their entirety. In these cases, this symptom will not appear to help in the diagnosis. Second, the relationship between customer and developer in addition to the developer's handling of methodology can make some requirements implicit. While this can work and thus save some time in creating and managing these extra requirements, it is important to be very careful about proceeding with this methodology. Misunderstanding will arise and test the strength of the relationship; therefore, the relationship must be able to weather these times or come to a poor end for one or more parties involved.

Too Many Options

Having some flexibility in the implementation of an application is important, and allows for improvements that would otherwise be impossible with purely rigid requirements. However, a programmer can easily find the number of options too abundant. Often in these circumstances, the developer will choose options based on some loose criteria that amounts to picking at random. Instead, it can be useful to look back at the requirements for the task and determine if there are any holes in the requirements.

This can easily lead to the discovery of missing requirements, which in turn can allow the programmer to collect these requirements and place them into the design. This provides the necessary added constraints to provide a much more solid basis for choosing among the possible implementations, and will lead to a greater chance of providing the needed functionality without performing several revisions.

While this symptom is useful for spotting unstated requirements, it does not usually become visible until it is time to implement the code. Spotting this symptom earlier can be achieved by considering the tasks at design time to get an idea if any of them have a large number of options. This can only be done if time is available, but keep in mind that details are not important at this stage. By considering only the general amount of possible options, without trying to choose among them or refine them in any way, the results can be achieved much faster at this stage.

History

Another useful technique for spotting missing requirements is related to the method used by regression testing. In regression testing, past results are compared against new results to ensure that no changes have been introduced that were not expected. Likewise, requirements for a new project can be compared against experience in past projects to ensure that the similar requirements have been covered in both. Obviously, just as intentional changes cause some regression testing to be ignored, differences in projects can cause some requirements to have no counterparts in the different projects.

Later in the section "Prevention," we will look at some of the most commonly overlooked requirements. The absence of these common requirements is a symptom of Requirement Deficiency that can be derived from the history of many projects throughout the history of software development. As you progress through different projects, you should encounter other common requirements that you might want to add to this list, reflecting your own particular history.

Nagging Feeling

Although not very methodical in nature, it is nonetheless useful to mention that intuition can play a part in discovering absent requirements. As was mentioned in the last section, learning from history is important to spotting the symptoms of

Requirement Deficiency. Not all of our development history will come to the forefront of our memories on demand, but might still provide guidance through the nagging feelings that our subconscious can generate.

This method of discovery is certainly not a replacement for more rigorous methods of requirements collection, but since it can only add new requirements that were missing, it deserves at least a modicum of attention and a small amount of time for investigation. It certainly would not be productive to spend weeks or months on a hunch, but an hour or two could lead to more information that confirms or denies the hunch in a more solid manner.

PREVENTION

Requirement Deficiency can be prevented by collecting all the necessary requirements, but how can it be determined that all the necessary requirements have been collected? The real work in preventing the existence of unstated requirements in a project lies in using techniques that minimize the chances of an important requirement being overlooked.

Communication

Clear and efficient communication channels are of great importance to getting all the requirements for an application. Although a developer does not necessarily need to be sitting next to an end user while he works, the more effectively the desires of the end user get communicated to the developer and the limitations of development get communicated to the end user, the more efficiently development can proceed, and the more usable of an end product will be generated. The major players in the chain that leads from end user to developer are end user, customer, management, and developer. Each of these players can contribute to the confusion of obtaining all the requirements; therefore, the more aware each is of his role, the better the requirements can be gathered (Figure 10.1).

The end user will be the final user of the product and will be very interested in the user interface and other usability issues. The chance for a developer to talk with an end user is very valuable, but even more valuable is the opportunity to observe the end user at work. This can reveal constraints and requirements of which the end user is not even consciously aware. The end user must also be encouraged by the

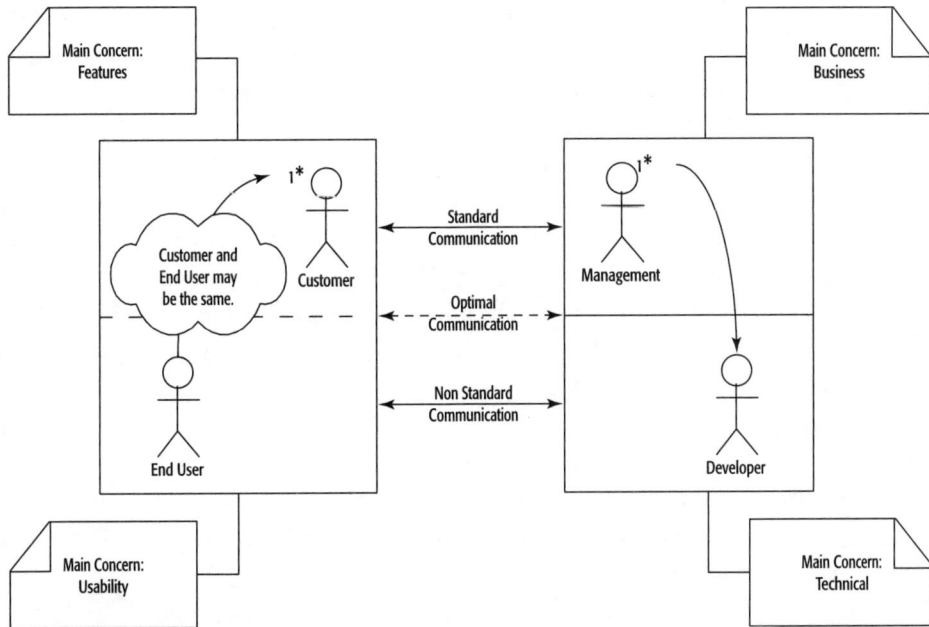

FIGURE 10.1 Communication and concerns of the standard parties involved in requirements and software development. Optimal communication involves all parties involved having direct communication with any other party when necessary, but it is more common and sometimes necessary to limit communication to the more standard channels.

developer to provide feedback whenever possible. One way to encourage this is to show the end user the benefit of his feedback when the next version is released. In shrink-wrapped software, a sampling of potential end users should be used during development to get an idea of the response the final end users will have when the product ships.

The customer, who might be anyone from the end user to someone at the end of a chain of managers above the end user, is most interested in getting an application that fulfills the needs of the corporation, and will often ignore the issues that concern the end user. The developer must encourage the customer to obtain this information, or the developer must be allowed to interact directly with the end users. The combination of the end user and customer requirements make up the core of the requirements of which the external parties are aware.

Management on both sides might have their own useful input on the requirements, but must above all else be encouraged to pass along information in a timely and complete manner. Revealing that a requirement exists late in the game, and was only unknown because it got caught up in management does more than damage the current project, it also damages the relationships of those involved and can lead to turnover or loss of business for the parties involved.

Finally, the developer who is programming and constructing the application will be most cognizant of the technical issues and limitations. The developer must be encouraged to communicate this information so that it can reach the customer and end user. This will allow the customer to make the necessary decisions about the importance of various requirements and other business-related dealings. The developer should also ensure that they have all the requirements they need, asking for more information if they feel anything is missing. This applies to individual programmers and their tasks as well as to the developer and application as a whole product.

The communication is two-way and it should be expected to last throughout the project in varying degrees. The efficiency of this communication is particularly important during the requirements phase as certain missing requirements are filled in and tradeoffs are worked out.

Paper Trail

While it is true that direct verbal communication is fastest and most efficient for working out issues, there is a need for a paper trail to exist for several reasons. Obviously, since we are talking about computers, "paper" is meant metaphorically. One of the most common methods of recording communications is through e-mail, but other forms of recording information are also useful. Documents, recordings, e-mails, and notes are all useful ways of recording the process and information transmitted during the requirements phase of development.

Perhaps the most useful aspect of recording these interactions is as a memory aid during development. Questions about the exact meaning of requirements or missing requirements might be answered in these archives without the need to revisit the entire process that occurred earlier. This can save time and ensure that the developer is clear on the exact requirements the application must meet.

Another important aspect of the paper trail is for settling disputes quickly and without undue impact on the schedule. This can range anywhere from a misunderstanding between programmers about their tasks to a disagreement between the customer and developer about what is expected for the application. This aspect of the paper trail will impact the way in which communications are recorded, requiring a more legal approach when customer and developer are involved and a less informal approach between coworkers.

Checklist

One of the symptoms of Requirement Deficiency is the lack of requirements that were there in previous projects. To facilitate this process, and therefore prevent these requirements from being overlooked, it is useful to keep a history of common requirements that have been encountered in past projects. This history can then be used as a checklist for ensuring that all the necessary requirements have been gathered for a new project.

There are several important nuances to keep in mind when creating this list. The first is that many of the useful items will appear later in development, not during the requirements collection phase. This follows from the fact that these requirements are not already on the checklist, and are therefore likely to be missed when the requirements are created. However, when these requirements are then discovered later with the associated problems, it becomes evident that they were missed and should be added to the checklist.

Next, note that you are compiling a general type of requirement that should be collected and not specific requirements. Therefore, you do not have a checklist that says:

- The application must perform at 60 frames per second.

Instead, you are looking for the more general:

- Performance requirements

It is also important not to add, or remove if discovered, items that are not common, and therefore not often missed from the checklist. Part of this process is deciding what type of project is generally associated with a particular requirement. Some requirements appear in almost every project, while others are numerous within a particular project type but unheard of in other types. What follows are some requirements that are common in many different types of projects.

Performance

Almost every application is concerned with performance to one degree or another. It is therefore important to obtain the performance requirements of the application. This can then be used as a guideline for generating a performance budget for different parts of the application, allowing for intelligent design decisions relating to performance. Be careful, however, that these requirements do not lead to Premature Optimization.

Memory Usage

There are still a large number of memory limited devices that are being developed for, and it is imperative that these limitations be included in the requirements. If an application becomes substantially larger than the target memory allows, missing this requirement can lead to a final implementation that will never be able to make it onto the device.

Even if the device is not severely memory limited, it is still important to know the memory of the target systems in order to avoid phenomena such as disk thrashing because of numerous page faults.

File Storage

File storage decreases in cost and increases in size every year, but file sizes also continue to increase in size, and seem to keep pace with available storage space. Be sure to consider file sizes, particularly if the application is meant for limited resource devices, or has files that will be commonly transferred over the Internet.

Stability

Every customer wants a bug-free application, but there is cost involved in testing and verification that cannot be overlooked. Be sure to get the requirements for what tests the application must pass to be acceptable, or you might find yourself with a lot of extra debugging that you did not anticipate. Requirements can range anywhere from provably bug free, a time-consuming effort that should only be used for life critical applications, to running for a reasonable amount of time without crashing and preserving data in the event of a crash.

Backward Compatibility

Many applications need to work with old data or communicate with legacy systems. Be sure the customer has given information on all the legacy products with which the application should be able to operate.

Extensibility

Customers would love a product that could be extended to meet any new demands they require, but reality limits the flexibility that can be added to an application, and remain cost effective. Obtain requirements from the customer about the areas in which the application is most likely to require extensibility. This is also a useful time to determine what areas are likely to change even before the project is finished.

Internal Technical Requirements

Customers are generally not knowledgeable about the technical information and limitations involved in implementing the application. Therefore, it is up to the developer to ask the right questions to obtain the information necessary for creating their own technical requirements. Obtaining these requirements is all part of the give-and-take process that was mentioned when we talked about communication.

First, the developer must acquire the general requirements the customer would like the application to fulfill. These should then be evaluated by the developer in order to generate the technical requirements. As this process occurs, some of the initial requirements might not be possible to meet, for technical, time, or budgetary reasons. These issues should be brought back to the customer after options have been outlined for alternatives or tradeoffs. This process might occur before the requirements are sufficiently refined.

A final step in the process should be to ensure that the individual developers agree that their part of the requirement can be accomplished. This feedback can help to further clarify vague requirements and add missing requirements. On the customer side, it might be useful to have the end user make any final suggestions concerning the tradeoffs and alternatives that were made to the requirements. As with so many parts of software development, a balance must be struck between the time taken to refine the requirements and the cost of the performing this refinement.

Flexibility

It is always important to maintain a certain amount of flexibility when developing software, but flexibility adds to development time and code complexity. This is where requirements can help focus effort, providing a guideline for when to provide extra flexibility. The best guide for adding flexibility comes from understanding the likely areas of change based on experience and customer concerns, but sometimes these are missing or poorly collected. Actually, it is where requirements are missing that demands the most flexibility. This is where the customer is most likely to decide later in development that the application must meet certain requirements.

Thus, by providing flexibility when the requirements are missing or vague, requests by the customer to impose these requirements later can be handled easily. This is of course assuming that you cannot get the customer to choose a stricter set of requirements early on. Since you cannot always choose who collects and provides the requirements, it is beneficial to leave flexibility where you are unable to obtain stricter requirements.

CURE

Once development gets past the requirements phase, there are still plenty of changes and decisions regarding these requirements. Many projects have gotten themselves in trouble because of the belief that the requirements would not change later in development. Some of these changes might have been accounted for in the requirements, but others might have been unstated and unknown to the developers.

Get Feedback

The earlier a missing requirement is identified, the easier it is to fix. To this end, the earlier the customer can provide feedback on the direction application development is taking, the better. The process of obtaining feedback can be a tricky business, involving the relations between the customer and the developer, but it is very important to avoid the scenario when the end product is not what the customer wants or will pay for.

For this, an incremental development cycle is the most useful type of methodology to use. Workable versions are produced on a regular basis for evaluation and feedback. It is important for the customer to understand that these versions are only part of the final application, but the responsibility for reaffirming this generally lies with the developer. Proper presentation of the application and knowledge of future development plans can assist greatly in assuring that there is no miscommunication or unfounded fears that come out of these feedback sessions.

Similarly, individual developers should look for feedback on their tasks to ensure that the result performs within the requirements. This can also lead to the discovery of missing requirements, most often because the writer of the requirements assumed that either there were no options or the option to choose was obvious. Peer reviews and proper testing can help enforce this early feedback within the team.

Incorporate

This feedback can then be incorporated into the requirements, allowing changes to be made on a timely basis rather than as a mad scramble at the end of development. Incorporating all the feedback into the requirements before acting on it is important because the changes might interact with each other and the current design in unexpected ways. Individual developers should provide input on changes that affect their individual tasks, preventing the new requirements from missing important information. Once again, good communication is the most important method for achieving this goal.

As each requirement is added, two main design changes might be required. The first design change requires modifications to the internal implementation, but not the application behavior, to accommodate the new requirements. The second type of design change requires additions that add new functionality or constraints to the final product. These will be most directly related to the customer's requirements, but cannot be accomplished until the internal changes are made.

Bargain

Due to technical, time, or budgetary limitations, some of the new requirements might not be feasible. It might be necessary to return to the customer with alternatives and tradeoffs if there are irresolvable conflicts. Be sure to have several options before doing this, allowing the customer to feel that they have a say in their final product. Remember that in addition to changing or removing features, it might also be appropriate to ask for more time or money to complete the new requirements if the customer must have them.

Individual programmers and developers should be particularly involved in determining if the new requirements are difficult or impossible because of time constraints or technical limitations. Most often, it is a question of time, but in essence, a large enough quantity of time amounts to the same trouble as technical unfeasibility. Once again, be sure that the communication channels remain clear and efficient.

Refactor

Unfortunately, the changes that might be required due to the new requirements can span the entire range of possible refactorings. One important guideline that can be followed, however, is to make any refactoring changes to the current code before proceeding to add the new code. This allows the proper unit and regression testing to ensure that the application's behavior does not change.

RELATED ILLNESSES

Improper handling of Requirement Deficiency can easily lead to a framework that is subject to the Brittle Bones illness. Even if the initial framework is not overly brittle for the original design, the changing requirements could weaken the structure to a point that it is almost unusable. Proper requirement collection can therefore assist in preventing both of these illnesses.

Although not necessarily caused by Myopia, both Myopia and Brittle Bones share a similar impact as Requirement Deficiency. All three are difficult to recognize in the short term, but have large detrimental effects over the long-term course of the project.

FIRST AID KIT

Ensuring that requirements are collected and remembered properly is assisted primarily by tools for proper documentation. There are many useful products out there for handling the creation of documents, spreadsheets, and other media for recording requirements. The most common is Microsoft Office, but other products such as Star Office from Sun and the open source Open Office provide alternatives that can still operate with most Microsoft Office files.

Storing and searching versions of requirements and related documents is a job for a good version control system. Perforce, Microsoft Visual SourceSafe, and CVS represent just a few version control systems that are available. Whichever system is chosen, it is even more important that with source code version control that the system be set up for ease of use. Requirements will be handled by not only programmers, but also managers and other team members who might not be experienced with version control systems.

Another important part of the documentation process that is often overlooked is e-mail. Information that was not officially recorded could easily have appeared in e-mail. There are a number of e-mail clients available, including Web browser-based e-mail. For documentation purposes, the Web browser clients, who commonly leave the e-mail on a remote server and therefore suffer some performance penalties, are less desirable than are clients that store mail on the local machine. Advanced searching features can be of particular importance when looking for information in e-mail, such as that found in The Bat e-mail client (Figure 10.2).

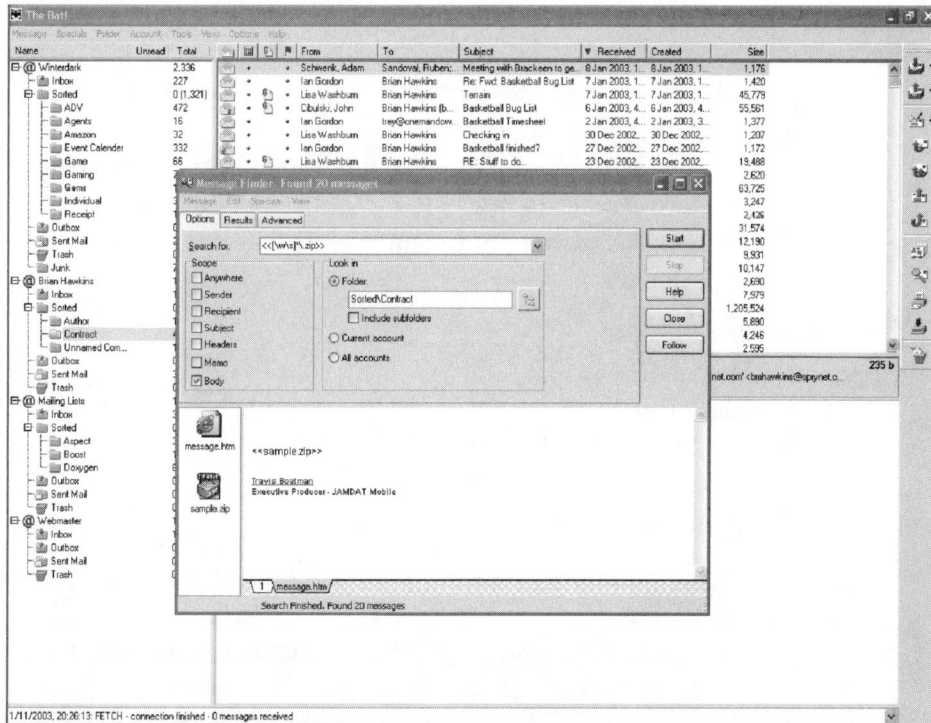

FIGURE 10.2 Advanced search features in The Bat allow the use of regular expressions to find specific information. Here a search is being performed for all messages that contain the text indicating a zip file attachment.

SUMMARY

The symptoms of Requirement Deficiency are:

- Vagueness exists in the requirements; in other words, the requirements lack the necessary level of detail.
- A lack of constraints on the implementation due to loose or missing requirements, leading to a plethora of options that must be chosen from in an arbitrary fashion.
- Requirements that have been in several past projects of the same or similar type are missing from the new requirements.
- Intuition that something is missing.

To prevent Requirement Deficiency:

- Establish clear and efficient communication channels all the way from the end user to the individual developer.
- Keep a record of communications about the requirements, allowing easy reference to more material in case of lost information or disputes.
- Make a checklist of common requirements that should be in every project, usually including:

  - Performance
  - Memory Usage
  - File Storage
  - Stability
  - Backward Compatibility
  - Extensibility

- Generate technical requirements, and use these to determine if changes or additions are required to the customer's requirements.
- Leave flexibility where the requirements are loose or missing to prepare for possible future demands in that area.

To cure the effects of Requirement Deficiency:

- Get feedback early and often, usually made possible by using an incremental development methodology.
- Incorporate feedback into the requirements as soon as it is received.
- Negotiate with the customer about alternatives and tradeoffs if the new requirements are not feasible within the time and budget allotted.
- Refactor current code before implementing new functionality associated with the new requirements.

11 | Myopia

DESCRIPTION

Shortsightedness plagues many human endeavors, and software development is no exception. If you have read the other illnesses up to this point, you might have noticed that each mentions Myopia as a related illness. Lack of long-term considerations permeates much of the poor reasoning that causes problems in development. This short-term approach causes immediate rewards to be visible to the programmer and his superiors, so it is not surprising that it continues to be encouraged. This insidious illness must be tackled with great care and patience if it is eventually to be made insignificant as a problem for software development.

SYMPTOMS

Because of its nature, the full-blown symptoms of Myopia are only manifested in the long term. However, subtler symptoms do appear that can help cut off the formation of these more serious symptoms.

Shortcuts

The number-one symptom of Myopia is the shortcut. Any time you find yourself thinking that you can take a shortcut to a solution, step back and consider the long-term consequences. More than half the time you will discover that your shortcut has a high risk of causing lost time rather than gained time. This is because shortcuts represent a short-term solution that only fixes the immediate problem. They do not consider future problems that might result either from not fixing the problem properly or from the shortcut itself.

PREVENTION

Preventing Myopia is not difficult to do, but it is difficult to start doing. Once you set yourself on the path of considering the long-term solution it becomes only a matter of execution. However, maintaining the discipline necessary to consider the long-term solution is the challenge to preventing Myopia.

Think Ahead

Above all else, the most important tool you have to prevent Myopia is your brain. It is not a bad thing to use instinct and educated guesses to track down a problem and come up with a possible solution. However, it is a very bad thing to implement that solution without first considering all the effects that the solution will have. This is where your brain steps in, and you must use it to consider what other possible effects a change might have.

When fixing a software problem, the goal is not only to fix the problem but also to have as little impact as possible on the rest of the implementation. This includes long-term effects on maintenance and possible changes. Too often, programmers rush to implement a solution that breaks other functionality. Even if this temptation is overcome, they are even less likely to consider such issues as readability, understandability, and maintainability.

Statistics

Many programmers start down the path because they desire to work in a more deterministic world that computers provide. Unfortunately, even though the execution of machine code by a microprocessor is deterministic, software development is certainly not that well defined. Software engineering is a young field with plenty of unanswered questions and constantly changing technology. The methodologies available are widely varied and relatively unproven. Compared to many industries, software development is chaos incarnate.

Once we have faced the fact that our chosen career is not the glorious utopia of determinism that we hoped for, is there any hope to reign in this chaos? There is in fact a solution, and it once again brings some reasonable and solid decision making back to the software development process. Instead of looking at development choices in a deterministic light, change the perception to one of statistical expectations. In other words, remember to take into account the probability of some future outcome in addition to the cost of that outcome. Without considering the probabilities, decisions will be made on an all-or-none basis that can lead to poor choices.

This is embodied in the concept of risk management. Assume that problems will occur and mistakes will be made. This makes the ultimate goal of development to reduce these errors rather than the vain attempt to eliminate them. This is accomplished by making decisions based on an analysis of the various outcomes each choice could have. By taking into account the probability of each outcome, and the estimated benefit or loss for each of these outcomes, a quantitative measure of the risk a choice has can be made. These measurements can then be used to compare choices on a more realistic basis.

The benefit of this approach is seen over the course of development. While problems will occur with the associated loss of time and money, more benefits will be seen on average that outweigh the losses. To better understand this, imagine that only a single factor of each choice were considered. For example, you might only consider the benefit obtained from each choice available. Now suppose that 75 percent of these failed, leaving only 25 percent of the benefit and much greater loss to outweigh this gain. On the other hand, if we chose the options that gave us an 80 percent success rate, then we are much more likely to come out with a positive result. Do not attempt to be perfect, attempt to minimize risk.

Education

Another important aspect of software development that programmers often forget after they leave college is continuing education. The computer industry is in a constant state of advancement, and commercial trends change quickly. New information and technology become available on a daily basis, and you must adapt if you are to keep an edge over the competition. The primary method for combating this ever-changing world of software development is continuing to learn and advance your personal knowledge.

There are many sources for furthering your education, and your choice of which to pursue will be based on your career and goals. However, the same basic formats that the information is available in are shared across most software disciplines. For example, if you are reading this book, then the chances are good that you understand the importance of books for continuing to learn. There are many other good books available, and consulting with fellow programmers can point you at some of the best.

Magazines are also a source of up-to-date information, and are often more timely than books. The information is somewhat more limited on any particular topic, but this is a fair trade for the more recent nature of the material discussed. Even more up to the minute are Web sites and newsgroups, although these must often be treated with a more critical eye. The speed with which electronic information is available is counterbalanced by the less strict evaluations that it undergoes. Magazines and books go through a more complete review process before publishing.

Sometimes printed material is just not good enough to get the information that you need to stay on top of your career. Conferences can be of great assistance in bringing together programmers who share similar interests. Conferences provide direct human interaction between people in the same industry and can be an excellent location to disseminate knowledge. Many have lectures, roundtables, and personal meetings available in addition to general gathering and sales pitches. One point of contention that often causes employers to be reluctant to send employees to conferences is the fear that they will only use the opportunity to look for another job. However, if this is the case, then the real question should be why they want to look for another job rather than how to stop them from finding one. Denying them the opportunity to go for further education is only going to further their desire to find work elsewhere. Therefore, if you are an employer you must not only consider your continuing education, but that of your employees as well.

Training courses represent another human interactive form of learning that is available for many of the tools of modern software developers. These are excellent opportunities to learn a product if you have never used it before. Unfortunately, the actual usefulness of these courses can vary widely. Finding other programmers who have attended one of the training courses for a particular product can help you gain a better notion of whether you should attend.

One of the best forms of learning can be found in the ancient practice of apprenticeship. Finding a good mentor can be an excellent way to learn more quickly with direct guidance that could not be found otherwise. Being a mentor can offer the opportunity for a new perspective that can help you stay fresh in the changing job of software engineering. Just remember that mentoring is two-way, both sides should learn from each other and be receptive to discussion. While a good mentor can greatly help you progress, a bad mentor can be an equally powerful hindrance.

CURE

You have fallen victim to Myopia, or are stuck with the result of someone else's Myopia. What do you do now? The first step is not to panic. You will need to step back and look at the situation with a clear mind in order to decide the best course of action.

Refactor

First, consider refactoring the current code to remove the problem introduced earlier. Refactoring is a natural part of software development, but in this case, it might involve some extra work that could have been avoided. Nevertheless, if it must be done, it can still be done in a controlled and orderly fashion to avoid introducing more problems instead of fixing the existing ones. Once again, the key is not to panic.

Decide what changes are required to get development back on track. Remember to account for any functionality that might be required later in development; otherwise, you will only be fixing a myopic problem with another myopic problem that will hit at an even more critical juncture in the project. Analyzing the problem and coming up with a solution is also a good break that can help to calm your fears so that you do not attempt to hack a solution.

Once a design is in place, the refactoring can begin. Refactor in small steps and test thoroughly as you proceed. This is always important in refactoring and prevents the introduction of more errors while you are attempting to solve old ones. Discipline is the key to avoid rushing in a poor solution, especially if deadlines loom. Rushing might allow you to meet that deadline, but it also increases the chances that you will miss the next deadline by an even greater margin.

Sometimes you really do not have the time to refactor and still meet all of your deadlines. This is the time when you must confront the decision on whether to give up features and simplify the goals of the project, or to extend the completion date of the project to keep all the features. As we just discussed, do not try to trade out the necessary time for refactoring just to keep a feature. If you do this, you are likely to lose more than just the one feature.

Reboot

Sometimes the result of Myopia is such a complete mess that it would take months to sort through and refactor the result. This is when you must step back and consider starting from scratch. Starting over might seem like it will only cost you more time, but when the alternative is reworking a piece of code that does not do what you want it to, it can be the best option.

This is always a judgment call that must be made on basis of an individual occurrence. You must estimate the amount of functionality that is similar to the desired functionality, and how much must change to meet the new requirements. You need to then estimate how long it would take to make these changes, keeping in mind that changing code requires more care and testing and therefore more time than writing new code. Now compare this to the time estimate for writing the functionality from scratch. This should give you an indication of which path to choose.

Avoid Crunching

There is another solution that comes up when a decision must be made between dropping features or pushing out the deadline when the result of Myopia is encountered, and that solution is to work long hours. Often known as "crunch time," this method of development has more disadvantages than advantages if used for more than a short period.

First, overworking can cause programmers to burn out or quit, which certainly does not help the effort to finish the project. Even if the programmers stay for the length of the project, chances are good that they will leave at the end of the project with all the knowledge and teamwork they have developed. This represents a loss that is often undervalued. The knowledge and skill of a good programmer is worth far more than technology.

However, if you are a programmer on a team and you must convince management that extended crunching is not a good idea, telling them that they will lose you is not the best first approach to convincing them. A better argument is that extending crunching does not work. The expected benefit of having programmers work extra hours is that the software will be completed faster. Unfortunately, the ratio of bug-free code to hours worked does not scale well for most programmers. As programmers work longer hours, they lose energy and concentration. Thus, they are more likely to introduce bugs as work goes on, and these bugs must then be fixed later before the implementation can be said to be complete. At some point, the time required to fix the bugs introduced in an exhausted state will exceed the time saved implementing new features. Crunching results in no gain or, even worse, in longer development times instead of shorter.

RELATED ILLNESSES

Myopia is listed as a minor illness because of its less substantial nature, but it is the most common of the minor illnesses. This is due to its participation in causing many of the other illnesses. This stems from the natural human tendency to be more influenced by the short-term circumstances rather than the long-term effects. For instance, Premature Optimization trades short-term performance gains for long-term schedule losses. Likewise, CAP Epidemic has its root in short-term coding speed gains while sacrificing long-term flexibility and maintenance costs. Although Myopia's link to NIH Syndrome is weaker, there are still many poor NIH Syndrome driven decisions that have their root in short-term thinking.

The list continues with Complexification and Over Simplification, which both result from localized programming decisions that do not consider the result upon the entire project in the long term. Docuphobia and i are similar to CAP Epidemic because they shorten coding time initially only to have later coding take substantially longer than the time saved. Hardcoding is almost entirely driven by myopic decision making, both in an attempt to save time and in the false assumption that designs are unlikely to change. Finally, Brittle Bones has domain over the initial stages of a project, which makes it an extremely likely candidate for Myopia to be a cause. As you can see, this minor illness has a tendency to appear in many places. Therefore, it is important to train yourself to think in terms of the eventual goals of a project in order to avoid falling into the trap of short-term thinking.

FIRST AID KIT

Preventing Myopia requires thinking ahead, and to accomplish this it is necessary to consider the entire scope of the project to some degree rather than fixating on only one aspect at a time. The best tools for that job are the ones that help present the project's scope to everyone working on the project. Project documentation is the most obvious means, and placing that documentation on a Web server can make it easily accessible. There are times when people might find it easier to use the documentation on paper, and for this, emerging concepts in data transformation using XML and XSLT can allow an organization to keep one copy of the documentation while formatting it appropriately for both Web deployment and print.

Another important tool that should be used on almost every project is the project scheduler. Tools such as Microsoft Project and Milestones can provide both assistance in creating the schedule and a multitude of options for presenting the schedule to all team members. As with all project documentation, this should be available in a form that can be easily reached by everyone, such as accessible through Web pages.

SUMMARY

The symptoms of Myopia are:

■ Use of shortcuts rather than the more difficult but more appropriate implementation, leading to greater costs later.

To prevent Myopia:

■ Consider future functionality and probable changes when determining the design for any part of the code.
■ Consider risk management factors in your decisions, trading initial development time for future savings.
■ Keep up to date about new software technologies and methodologies, trading learning time for improved development techniques that save more time in development.

To cure the effects of Myopia:

■ When shortcuts and hacks are taken in the code, either by mistake or to meet a short-term deadline, then the code should be refactored to the more appropriate solution as soon as possible.
■ If refactoring is complex and time consuming, consider the option of starting from the beginning again.
■ Avoid crunch time as much as possible.

12 | Conclusion

Whether you have read through each individual illness or just a few that sounded interesting to you, you now have the techniques and information necessary to improve your personal programming abilities. There are certainly more illnesses to be discovered and analyzed, but before we delve into the future of preventing and curing programmer illnesses, we will examine several common concepts and techniques that can help with most types of illness.

Common Techniques

The following concepts and techniques can help in the prevention and cure of almost every programmer illness.

Language: The Gilded Cage

gild·ed \'gil-dəd\ [ME *gilden*, fr. OE *gyldan*; akin to OE *gold* gold] … 2 … **b:** to give an attractive but often deceptive appearance to …

— *The Merriam Webster Collegiate Dictionary*, Merriam Webster, 1997.

Language permeates our lives and allows us to communicate with each other and our descendants. Circumstances and history have created many different human languages to deal with the needs of people who use the language. The complexity of our languages is one of the most important contributors to the advances the human race has been able to make.

Nevertheless, language is not without its faults. Many tend to forget or overlook the fact that to go from thought to speech forces us to apply the constraints of the language we are speaking to the thoughts we want to voice. This traps our ideas in a cage of our own making, and we can only break them free through the evolution of our language to handle new concepts.

Take for example the Chinese word *chi*, which has equivalent words in several other Asian languages, that is summarized by the following definition:

> *Chi (also qi, ch'i, or ki): A subtle internal life force or energy. This energy manifests it-self as work, or a force that allows something to function. Chinese, Japanese, and Korean cultures are permeated with this concept, which breaks the perception of the world into physical matter and invisible energy.*

First, even if this definition were sufficient, it would still be a lengthy replacement for the single word *chi* used in the Chinese language. However, notice that the definition mentions that the concept of chi underlies Chinese culture. This word is directly tied into the culture and history of the Chinese people and therefore has a much richer meaning that cannot be fully translated into the English language. The true meaning and purpose of the word is therefore inseparable from the system of which it originated.

This connection between language and the system from which it originates also acts as feedback that reinforces the culture that generated it. Part of the difficulty of understanding the word *chi* for English speaking people is the lack of cultural context and corresponding rigid thought patterns that prevent the understanding of the full meaning . These concepts about language can be applied not only to human language but also to programming languages.

Programming languages provide similar benefits and suffer from similar faults as human language. Programming languages allow us to communicate with computers and other devices containing microprocessors in a manner that would otherwise be impossible. Without them, constructing software for even the simplest devices would be so time consuming that it would not be a viable business except for a few special applications.

However, programming languages suffer from even more constraints upon the ideas we want to communicate to the computer. Due to the more rigid requirements that computers impose upon communication, languages must apply tight constraints and policies upon communication if the device is to understand the information. The chain of communication therefore goes from the domain of the problem to be solved, then to the human programmers who must communicate the problem and solution ideas using human language, to the implementation in the programming language that the computer requires to understand and perform the requested problem solving. This chain requires at least two translations to be performed from the initial domain, with the ensuing constraints that this implies. It is not difficult to see from this that much can be lost on the path to developing software.

The constraints of the language are not the only problems that translation between domain and languages causes; they also introduce the possibility of errors. The likelihood of mistakes increases with the complexity of the translation that is required. Some of these mistakes are simple human error that is unavoidable with the degree of complexity of the translation, but others are caused by a lack of understanding of the programming language by the programmer. This lack of understanding can easily lead the programmer to other solutions that are more complex and therefore more error prone.

Over time, the constraints of the language will affect the problem-solving techniques that the programmer uses. This trend further constrains that programmer to the solutions possible in the language that predominates his programming experiences. Even as the programming language evolves or new programming languages are adopted, the temptation to fall back on known practices is hard to resist even if these practices cause numerous problems.

What this means to you as a programmer is that there are two main competing forces in the decision of what language to use for a particular problem. The first factor is which language is most appropriate for solving the problem. This is determined primarily by the complexity of translating the problem from human language and design representation to implementation in the programming language. The simpler this translation is, the more time that is gained in both the translation and the lower number of errors that must be fixed.

The second factor is the level of competence of the programmer with the language that is chosen. Even if the translation is simpler for a particular language, if the programmer is unable to make the simpler translation there will be no benefit to using that language. In fact, it might be a detriment as the programmer struggles with the language and introduces errors that would not have happened in a language with which the programmer was more comfortable. This will most often occur when a programmer is delegated a problem and the language to use by a manager, because if the programmer performing the implementation were to choose the language, then the knowledge of the simpler translation would not factor into the decision.

Throughout most of the illnesses, the main techniques were kept as language independent as possible to encourage the use of the appropriate language for the given task with the programmers available. To fully take advantage of multiple programming languages, new languages must be learned and explored by both the individual programmer and the colleagues with whom the programmer works. This knowledge provides an extra benefit aside from opening up the number of languages available for solving a problem. Understanding multiple languages can also loosen the constraints built up around the problem-solving mechanisms introduced by a particular language. This can even lead to discovering that some techniques from a new language translate well to a language already in use by the programmer. These techniques might have simply been overlooked by the most common problem-solving paradigm used by programmers of the older language.

Risk Management

Despite our desire to make it so, software development will never be a deterministic process. Mistakes, unexpected events, and other outside forces will work to prevent any plan from completing as written. We all know this, yet so often we choose to ignore this knowledge in favor of a plan that looks optimal on paper. The problem with the optimal plan for any software project is that it often requires everything to go as expected. When a part of the plan fails, the intricate dependencies that are required for the plan to be optimal fall apart, and drag the project down into a quagmire from which it might never recover.

Instead of creating a plan that rests on the fine edge, create a plan that is meant to minimize the risks involved in mistakes and changes. This does not mean to play it safe, and only develop products that have been done before and are guaranteed to succeed. These projects are just as likely to encounter random events that cause failure, or they can simply stagnate as others pass the old technology by with more advanced techniques. Managing a project risk is balancing the risks against the benefits, on occasion choosing riskier plans if the payoff is worth it and sticking to the safer plans when the risky plan would not pay off sufficiently.

Testing

One of the most misunderstood and misused of the software engineering techniques is testing. The two primary misuses of testing are too much or none at all. Programmers will often swing from one of these extremes to the other. Perhaps the last project was a debugging nightmare as last-minute changes broke the build in numerous ways that were difficult to track. This project, everything will be tested down to the finest detail. On the other hand, maybe the last project was so bogged down by testing that everything was delivered late or with missing features. This project does not need the extra overhead that burdened the last project.

Of course, if the last project was completed, then you might just decide to keep the status quo. After all, the other extreme only changes the problems encountered without reducing them. This line of thinking often arises from talking with programmers who have worked on the other side of the extreme. If it is not broken, do not fix it.

However, the process is broken. Countless hours are wasted debugging on the one side, and countless hours are wasted writing unused test cases on the other side. Therefore, we should search for a solution to this situation that will retrieve these lost hours. Consider the optimal amount of testing that a project could hope to achieve. A test case would be written for each instance where the code is broken upon writing, or broken in the future, and only for these instances. Unless you have a crystal ball tucked away in your office and the knowledge to make it work, achieving this optimal solution is unlikely to happen.

Instead, we again bring into focus the ideas of risk management. The value of a test case is determined by the amount of time that the test would take to create versus the amount of time that is lost in debugging and other testing when the test case is not created. Obviously, neither of these values is known when the decision to write a test must be made. You must therefore estimate what these values would be based on past experience and your knowledge of the workings of the code.

Estimating the amount of time that the test will take to write is the easier of the two tasks, and is identical to estimating the time it would take you to implement any well-defined piece of code. The more difficult estimation is the amount of time that will be lost to debugging. This value is made up of two parts, the amount of time that a failure would take to analyze and fix, and the probability that another programmer will break the code in the future. In fact, you might even be the one to break the code later, but it is best not to consider this because it is more likely to bias your answer toward too little testing. Weighting the debugging time with the probability of failure gives the appropriate value to consider against the time to write the test. This idea can be more succinctly expressed by the following rule:

■ *Tests should be written to catch the most likely points of major failure.*

One particular software methodology in which this rule can be of great assistance in balancing the time spent writing tests with the benefit of the tests is known as *test-driven development*. In this methodology, the tests are written first, and then the code to make the tests work is added. This code is added first with just enough to allow the test to run and fail. This proves that the test can fail. Then the correct functionality is added in until the test completely passes. Writing the tests with consideration to the most likely points of failure provides twice the benefit for minimal work because the failure cases will be guarded against both later and through the initial development of the tested code. This initial safeguarding does not happen when the test is written after the code.

However, even if you use this methodology, do not ignore the need to write tests after the code is already developed. When an error occurs, write a test for it because errors tend to happen in the same place more often than they do in new places. This new test can then correctly safeguard against this error in the future, even though it was not available to save the debugging labor the first time through the coding.

How to Test

Once you are convinced that testing is an important part of the software development process, you will become curious about what to test and how to test it. There is plenty of literature available, although much of the literature is for large projects or full testing departments. Here we are more concerned about testing as it applies to what an individual programmer can accomplish. A good set of guidelines on testing can be found in [Hunt00]. To place this in context with the prevention and curing of illness, let us look at a summary of the testing process as it applies to one or only a few programmers.

Testing types can be divided into three main categories:

- Unit testing
- Integration testing
- System testing

This is a simplification of the types given in [Hunt00] so we can look at the applicability of these types to the individual programmer.

Unit testing is the testing of a single code unit such as a module, class, or file. This level is best handled by one or two programmers familiar with the code and its goals. Having a second programmer, who did not write the code to be tested, write the tests can be an excellent way to prevent the common habit of writing tests that subconsciously avoid the breaking points of the code. Writing the tests before writing the code is another approach to help prevent writing tests that avoid the points of failure. The tests are directed at the responsibilities of the single unit. To aid in the maintenance and applicability of the tests they should be kept with the code they are testing, in either the same file or directory. Naming conventions that place test as a prefix or suffix to test cases, and test cases only can also be useful, particularly for automated testing and other automated tools.

Integration testing is the testing of the interactions between separate modules, classes, or files. This level involves anywhere from a single programmer to the entire team, but is still generally the responsibility of an individual programmer to actually write the tests. It is important to focus on the interactions between the modules and avoid regressing to unit testing that might only focus on an individual feature from only one of the modules.

System testing is the testing of the entire application, including its usability and adherence to the customer's requirements and expectations. This form of testing is most often carried out by a person or group other than the programmers, and therefore only the results are of primary concern to the programmers. However, it is still useful to know the basics of the testing techniques being used in order to reveal flaws in the testing or the reporting of results. This can aid you in the end by providing better information at an earlier date to eliminate errors from the application. Many of these tests are not written in code, especially given the need to test and evaluate the user interface and other customer relevant factors.

When testing at the unit and integration level, the goal and therefore method of testing can again be divided into three primary types, which could also be thought of as the three Rs for easy recall:

- Result testing
- Resource testing
- Regression testing

Result testing is the first testing that is generally written, and aims to test that the unit of code achieves the goals it was created to reach. This form of testing should also verify that the code handles errors as expected, once again given the proper result according to the requirements. Passing this test provides the necessary condition to certify that the code is working as expected. However, this does not necessarily indicate that the code is complete.

Resource testing tests whether the code performs within the limits of the resources available to it. This includes testing performance, memory usage, and I/O handling. Resource testing should perform these tests both under normal operating circumstances and under heavy load. In the case of heavy load, the code should continue to either operate or fail gracefully. Performance testing tends to be the most heavily emphasized of this type of testing, but the other tests should receive sufficient attention. This is particularly important under certain architectures.

Finally, *regression testing* occurs after a unit of code has passed result testing. The results from previous tests can be used to verify that the code continues to operate as expected. This is particularly useful for refactoring, where the goal is to improve code readability and maintainability without affecting the results. It is also very important for performance and other resource tuning, as their goal is also to change only the resource usage of the code without affecting the results.

This is only a brief overview of the process of testing, and you are encouraged to find out more about testing. There are libraries available for most of the common programming languages to assist in testing. For example, Java has JUnit, and C++ has CppTest. Using these libraries is important because it removes part of the tedious and error-prone part of testing, thus making testing much more useful. Regardless, testing is extremely important in whatever manner you choose to perform it.

Refactoring

Refactoring is another important technique that is essential to enhancing your skills as a programmer. Without the proper knowledge of refactoring, it becomes easy to spend too much time in the design stage trying to compensate for every possible change. The other option is to give up on the design entirely and end up with chaos as design changes are requested later in the project. To avoid these fates it is important to understand the concept of refactoring as well as learning and creating techniques for easier refactoring.

The ideas and benefits behind refactoring are best described in [Fowler00]. This book not only outlines the concepts of refactoring, but also provides a large number of systematic instructions and examples for performing many common refactorings. With this excellent reference already available, there is no reason to cover refactoring in detail. Instead, a brief description of the essence of refactoring and why it is important in the prevention, and even more so in the curing, of programmer illness gives the important factors related to the current discussion.

The basic idea of refactoring is to provide a systematic means of making changes to existing code for the purposes of fixing errors, understanding the code, and adapting to changing requirements. Programming illnesses are a major source of errors that can be fixed by proper refactoring. The benefit gained by fixing these problems in a more orderly fashion is the prevention of further errors caused by mistakes made while attempting to perform the required adjustments. In addition, proper refactoring can also improve the readability and maintainability of the code. This makes future changes and error fixes easier to accomplish.

Refactoring also provides enough confidence in the ability of the programmer to adapt to change so that the programmer can move forward once a reasonable level of design is achieved. This prevents analysis paralysis, or the tendency to become stuck in the analysis and design stage trying to anticipate all possible outcomes. However, a word of warning is important here. While a small disciplined team can achieve major success with minimal design and refactoring as advocated by Extreme Programming, not all projects are suitable for this approach. Large teams or less skilled programmers require a more complete design, even if they understand the basics of refactoring. This reflects the complexity of interactions on large projects or less skill in handling the interactions. Therefore, use refactoring as a tool to prevent over design and to fix errors, but do not use it as a crutch to excuse poor design and sloppy coding.

Toward the Programmer's DSM-IV

In 1952, the American Psychiatric Association published the first edition of the *Diagnostic and Statistical Manual of Mental Disorders* (DSM-I), based on a classification system for mental illnesses created by Emil Kraeplin (1856–1926). The hope was that the disorders could be grouped to determine the proper treatment and understand the course of the disorder if left untreated. Unfortunately, the DSM-I did not fully live up to these expectations. However, through a series of revisions in 1968 (DSM-II), 1980 (DSM-III), 1987 (DSM-III-R), and 1993 (DSM-IV), the reliability and validity of the manual has increased toward meeting these expectations. [Sue94]

A similar aim can be seen for books and discussions about the mental oddities that cause so many problems for programmers of all types. Just as with the DSM-IV, the goal of this book is to improve the classification and treatment of common programming problems caused by incorrect or misinformed thinking by programmers. Also similar, there is an attempt to stress that there are several aspects to these problems, which include the project environment in addition to the thought processes of the programmer. Although not as clinically laid out as the five-axis evaluation method of the DSM-IV, there is still an attempt to stress the importance of the environment and the interrelationship of the various illnesses when it comes to diagnosing, preventing, and curing programming illnesses.

This book represents only a starting point for a more solid classification of the problems that programmers experience and a structured approach to curing these problems. This is not the first book to approach software development from the point of view of common problems and their solutions [Brown98], but it does provide an emphasis on the mistakes made by the individual programmer and a more complete approach to the thinking behind these mistakes. This is necessary to overcome the cause of the problem rather than continually fixing the results. Just as with the DSM, the reliability of the diagnosis of symptoms must be confirmed or refined. The validity of the proposed solutions must also be confirmed or improved to meet the needs of all programmers.

Community

In order to further both your knowledge of programming and the overall collection of programming knowledge, it is important to participate in the software development community. This will allow the industry as a whole to continue to grow and become one that can last for a long time to come. Discussion about the topics in this book is only one of the many venues that need exploration and discussion from programmers in all corners of the world.

There are a plethora of ways to learn and contribute to the software development community, including books, magazines, mailing lists, newsgroups, Web sites, and conferences. Just as you might find answers to questions that would otherwise take you a great deal of time to puzzle through, you might also be able to quickly answer questions that others are having difficulty solving. Most importantly, contributing the knowledge that can only be gained by experience allows the overall knowledge of the community to increase in ways that would not be possible without open communication.

A common objection that comes up when discussing participation in the community is proprietary information. This is a valid concern, although not to the degree that it is often taken. It is, however, unlikely that everything you do will fall under the umbrella of secret information. Much of the code that is written to support an application is easy to recreate and does not represent a business advantage. The primary advantage comes from the integration of the separate parts into a whole package, along with proprietary assets, and a few key sections of code that contain proprietary algorithms. There is generally still plenty of knowledge and techniques gained that can be shared without giving up the competitive advantage, and in return, this encourages others to contribute information that you can use to increase your development speed and quality.

APPENDIX
A Teamwork

Up until now, we were primarily focused on the individual programmer and techniques he could use to improve his performance. There was less discussion of the interaction between programmers. Because the majority of programmers will find themselves working in a team environment, it is important that programmers be able to interact with other programmers, other members of the developer's organization, and in some cases, the customers and end users of the product. This short introduction to such an extensive topic is intended to just get you started thinking about the effects of team dynamics on software development.

Made for People

While deep in development, it is easy to forget that the final product will be used by other people. It is also likely that the users will not be programmers, which means there are several considerations that can easily be forgotten. Communication to the customer is a key area from the very beginning of development. The average customer will not understand the technical issues involved in what they are asking for in a product, leaving it up to the developer to provide a basic understanding of the limitations that might be encountered during development.

Customers will not like to hear that their requested features cannot be implemented, and can easily turn to another developer if they feel the reason for this failure is a team's incompetence. To combat this possibility, it is important to have a list of alternatives available when presenting the limitations. Alternatives can include similar features that are substantially easier to implement, or budget and time changes that would provide the necessary resources to implement the feature.

This communication is not limited to the beginning of development, and becomes particularly important if unexpected problems arise later in development that place a feature in jeopardy of not being implemented. Customers will respond much better to clear explanations and a variety of choices than they will to stories of woe and confusing technical explanations.

Buffering the communication between customer and developer with a layer of management on the developer's side can be either a benefit or detriment, depending on the circumstances. If the manager is both technically knowledgeable and capable of translating the technical details into easy-to-understand terms, the situation is optimal. Otherwise, it is still up to the technical developer reporting to the manager to place the explanation in understandable terms. However, even if the manager is not technically knowledgeable, he can concentrate more on communicating with the customer to build a strong rapport that will make the exchange easier. At some point, though, there will need to be a translation from technical terms into layman terms.

The other major consideration is how the customer affects what choices are made during implementation. In a manner of speaking, the dialogue between the customer and developer continues even after the product is finished in the form of the end user using the product. The end user, and therefore the customer, is only concerned about the elements that he can see and hear. A slick implementation of a particular algorithm is only useful to the customer if the effect is visible on the final product. Thus, one of the considerations in choosing an implementation is the visible effect it will have on the product in the customer's eyes. However, do not forget that development time is also visible to the customer, so implementation decisions meant to decrease development time are also very valuable.

The main point is that the goal of the product is to impress the customer and end user. While nifty algorithms might impress fellow programmers, and cool technical terms can benefit marketing somewhat, an application that meets the user's needs effectively and efficiently will provide the most impact. Except for academics, most programmers must remember that they are part of a business. Even for academics there is often politics and funding to compete for, and these require some of the same attention to the target audience that is given to the customer in commercial products.

Made by People

Another fact often overlooked by developers is that people are making the software. A developer is not a black box that takes a list of requirements and magically produces the product. Humans, with all their quirks and oddities, are a key part of the process. Understanding this is a key part of avoiding problems and delays caused by the personalities involved in development.

The most important factor that the human involvement brings to software development is the need for communication. Even with the advanced capabilities of human language, there is still plenty of opportunity for communication failure to cause problems. Knowledge of these common failures can help avoid them, just as knowledge of common programming mistakes allows those mistakes to be prevented. Shortly, we will look at the importance of different viewpoints on communication, but for now let us look at the other effects of communication on development.

Apart from the point of view that affects the communications of team members, other human biases will also affect what and how people choose to communicate information. Before assuming that another team member is stupid or stubborn, consider how his personal bias and career history might be affecting his decisions. Not only will this reduce your level of frustration, it will allow you to prepare a better argument to change his mind if you still believe he is wrong.

Just as the customer and developer can have different technical knowledge, so can individuals within a developer. Some people are better than others at communicating between different groups, and understanding who they are on a team can avoid confusion. Both sides, technical and non-technical team members, can take advantage of these people in order to aid communication within the developer. This will reduce frustration that typically comes when two people with disparate technical experience attempt to communicate technical issues.

If the developer is a large organization, another danger is the number of people involved in communicating any one piece of information. The more people the information travels through, the more likely it is to be distorted or misunderstood. In addition, response time is reduced, sometimes to the point where vital questions are not asked because of the difficulty in getting timely answers. E-mail and other permanent forms of documentation can help reduce communication errors. Whenever possible, passing along the original communication unchanged, along with any annotations, is best for ensuring that the destination party fully understands what the source of the communication meant. Response time is improved by determining which parties require direct communication on a regular and timely basis. This communication should be established and encouraged unless there are personality issues involved that would cost more time than the faster communication channel saves.

Another important consideration in working with people is the need to prepare for occasional mistakes and poor decisions. Even the best programmers introduce bugs into code and choose the wrong path on occasion. The difference is in how well prepared the team is to detect and handle these errors. Proper testing, peer reviews, and a willingness to consider oneself fallible go a long way toward minimizing the impact of errors. In other words, do not attempt to eliminate all errors; rather, attempt to minimize their impact.

Point of View

Different members of a team have different, sometimes opposing, stakes in the outcome of the project. Understanding the point of view of other team members can assist in both communication and decision-making.

Management

Programmers have the most difficulty understanding management. The motivations of the managers, which are primarily focused on business decisions, often differ from those of the programmers and other developers, which are primarily focused on the technology and creative aspects of development.

In order to be persuasive when talking to management, a programmer must remember these motivations and present his ideas in terms of these motivations. For example, when suggesting the purchase of a third-party library, show how the schedule can be reduced to save more money than will be spent on the library. If a feature cannot be reasonably completed, explain how much extra time and money would be required to complete it. In addition, offer alternatives that might provide a reasonable replacement at much less cost and time. Do not be afraid to ask which areas are most flexible for the customer and management: time, cost, or features.

Likewise, look at requests from management in terms of the customer and business concerns. Some features might sound meaningless, but the customer might have some reason for the feature and management is asking you to implement the feature to keep the customer satisfied. Nothing useful is accomplished by returning with a refusal to add the feature. A better approach is to inform management of the costs of implementing the feature and any tradeoffs that will be required. If they are willing to accept those costs, but you still have reasons for believing the feature is detrimental to the product, present your concerns as concerns for the product and suggest alternatives that will still satisfy the customer.

As always, keep in mind that you should think about the request first, but everyone can make mistakes. If you believe a mistake has been made, present your arguments rather than refusing without a good explanation. This might seem obvious, but it is easier to forget to do this than most people think.

Team Members

On the other side of the equation, it is important for management to understand the point of view of the individual programmers and other individual developers. To many of these individuals, the technical merit and correctness of the application will be more important than esoteric business concerns that they might not even know exist. Some of this can be overcome by explaining the business reasons for a request. With a better understanding of the reasoning, you might find that some programmers become less reluctant to perform operations that go against their first instincts.

When the reasons still do not convince the programmer, be sure to listen to his objections and get feedback on the impact to both the application and the schedule of the requested feature or change. In addition, listen to any alternatives. Once you have this feedback, you are in a much better position to decide on these alternatives or explain to the programmer that you are willing to accept the tradeoffs.

Coworker

Finally, do not forget that even your coworkers have different ideas and motivations than you do. If you do not understand why they are implementing a piece of code the way they are, ask them for an explanation. In so doing, you might learn new concepts and techniques, or you might have an opportunity to point out concepts or techniques the other programmer did not consider. Thus, instead of wasting time and becoming frustrated, the project will progress smoothly and ideas will be shared.

This is of course the ideal, but reality must occasionally disrupt this programming utopia. There are programmers who are not capable of the proper level of performance to be a member of a particular team. There are also programmers who are disruptive to the morale of a team. While a certain amount of patience and tolerance is essential to a team environment, managers must realize that some relationships will not work, and the relationship must therefore be ended. As with so many other aspects of software development, and life in general, maintaining an efficient and skilled team is a matter of balancing the different forces at work.

APPENDIX B
References

[Alexandrescu01] Alexandrescu, Andrei, *Modern C++ Design*, Addison-Wesley, 2001.

[Bentley98] Bentley, Jon, and Bob Sedgewick, "Ternary Search Trees," *Dr. Dobb's Journal*, April 1998.

[Brooks95] Brooks, Jr., Frederick P., *The Mythical Man-Month*, Addison-Wesley, 1995.

[Brown98] Brown, William J., Raphael C. Malveau, Hays W. McCormick III, and Thomas J. Mowbray, *AntiPatterns: Refactoring Software, Architectures, and Projects in Crisis*, John Wiley & Sons, Inc., 1998.

[Czarnecki00] Czarnecki, Krzystof, and Ulrich W. Eisenecker, *Generative Programming: Methods, Tools, and Applications*, Addison-Wesley, 2000.

[Fowler00] Fowler, Martin, *Refactoring: Improving the Design of Existing Code*, Addison-Wesley, 2000.

[Frantzis98] Frantzis, Bruce K., *The Power of Internal Martial Arts*, North Atlantic Books, 1998.

[Freidl02] Freidl, Jeffrey E. F., *Mastering Regular Expressions*, O'Reilly & Associates, 2002.

[GoF95] Gamma, Erich, Richard Helm, Ralph Johnson, and John Vlissides, *Design Patterns: Objects of Reusable Object-Oriented Software*, Addison-Wesley, 1995.

[Harris74] Harris, Marvin, *Cows, Pigs, Wars, and Witches: The Riddles of Culture*, Random House, Inc., 1974.

[Hunt00] Hunt, Andrew, and David Thomas, *The Pragmatic Programmer*, Addison-Wesley, 2000.

[Levine92] Levine, John R., Tony Mason, and Doug Brown, *lex & yacc*, O'Reilly & Associates, 1992.

[Reynolds87] Reynolds, Craig W., "Flocks, Herds, Schools: A Distributed Behavioral Model," Computer Graphics 21(4), July 1987.

[Sedgewick90] Sedgewick, Robert, *Algorithms in C*, Addison-Wesley, 1990.

[Sklansky87] Sklansky, David, *The Theory of Poker*, Two Plus Two Publishing, 1987.

[Sue94] Sue, David, Derald Sue, and Stanley Sue, *Understanding Abnormal Behavior Fourth Edition*, Houghton Mifflin Company, 1994.

About the CD-ROM

This CD-ROM contains source code and software to assist you in improving your programming techniques. The CD-ROM is organized into two main folders, one called Source containing the source code examples from the book, and another called Tools containing software that is useful for improving programming efficiency. In addition, copies of all the figures from the book are provided in the Figures folder.

Source

The Source folder contains both C++ and Java examples. The C++ examples are available under the Examples folder and are further organized by chapter. For convenience, a Microsoft Visual Studio .NET project file is provided.

The Java examples can be found under the JavaExamples folder under the com.crm.ppt.examples package. The Java examples are also organized by chapter. For convenience, an Apache Ant build file is provided.

Documentation for all examples can be found in the html folder.

Tools

Several tools are included on the CD-ROM; some are full versions while others are demonstration versions. To check for newer versions or implementations for other platforms, please visit the Web site for the tool. Each tool is in its own folder under the Tools folder:

- **Cmake:** Platform-independent build utility for C/C++ inspired by the Apache Ant build utility for Java. Please read the accompanying license. Available as a binary Windows install or source code in either zip or tar/gzip format.
- **CppUnit:** Testing library for C++, similar to JUnit for Java. Provides utilities for writing automated testing, including an accompanying GUI interface. This

library is covered under the GNU Lesser General Public License. Available as a single zip file or two Unix tar/gzip files for source and documentation.

- **Doc-O-Matic:** Automated code documentation without the need for special comment formatting. Particularly useful for legacy code. Demonstration version available as a Windows install.

- **Doxygen:** Automated documentation utility for C/C++, similar to JavaDoc for Java. Requires special commenting format. This software is covered under the GNU General Public License. Available as a binary Windows install or source code in either zip or tar/gzip format.

- **Milestones:** Project management and scheduling software. Available as a Windows install in either exe or msi format.

- **Perforce:** Version control software that provides a balance between price and features. Full-featured version for two users only available as Windows install. Contact Perforce for information on licensing more users.

- **Ruby:** Fully object-oriented language particularly useful for scripting and fast prototyping. This software is covered under the GNU General Public License. Available as a binary Windows install.

- **Visual Assist:** Enhanced editor plug-in for Microsoft Visual Studio. Provides auto-completion, code templates, and other useful features to make code editing more efficient. Available as a Windows install for Microsoft Visual Studio 6 or Microsoft Visual Studio .NET.

- **WinCVS:** GUI front end for the CVS version control system. Not as easy to use as other version control systems, but this is balanced by the fact that it is free. This software is covered under the GNU General Public License. Available as a binary Windows install.

System Requirements

The source code is intended to be standard C++ and Java code. It has been tested using Microsoft Visual Studio .NET compiler for the C++ source code and the Sun JDK 1.4 compiler for the Java code.

The software utilities provided require Windows 98, 2000, NT, or XP Professional. Although most of the software demos work with Windows 98, to use the Microsoft Visual C++ .NET project file you need Microsoft Visual C++ .NET which requires NT, 2000, or XP. Some programs might have additional requirements, so please refer to specific programs for their information.

Index